AF412959

Shlomo Melmed   Henri Rochefort
Philippe Chanson   Yves Christen (Eds.)

Hormonal Control of Cell Cycle

With 31 Figures and 1 Table

 Springer

Shlomo Melmed, M.D.
Cedars-Sinai Medical Center
8700 Beverly Blvd. Room 2015
Los Angeles, CA 90048
USA
e-mail: Shlomo.melmed@cshs.org

Philippe Chanson, M.D.
Service d'Endocrinologie
Hôpital Bicêtre
78, avenue de Géneral Leclerc
94275 Kremlin-Bicêtre
France
e-mail: philippe.chanson@
bct.ap-hop-paris.fr

Henri Rochefort, Ph.D.
CRLC Val d'Aurelle
Bat B Cancer Research Center
34298 Monpellier Cedex 5
France
e-mail: henri.rochefort@valdorel.fnclcc.fr

Yves Christen, Ph.D.
Fondation IPSEN
Pour la Recherche Thérapeutique
24, rue Erlanger
75781 Paris Cedex 16
France
e-mail: yves.christen@ipsen.com

ISSN 1861-2253
ISBN 978-3-540-73854-1

Cataloging-in-Publication Data applied for Bibliographic information published by Die Deutsche Bibliothek
Die Deutsche Bibliothek lists this publication in the Deutsche Nationalbibliografie; detailed bibliographic data is available in the Internet at <http://dnb.ddb.de>.

Springer is a part of Springer Science+Business Media
springer.com

Cover design: *design & production*, Heidelberg, Germany
Typesetting and production: LE-TEX Jelonek, Schmidt & Vöckler GbR, Leipzig, Germany
Printed on acid-free paper   27/3100/YL   5 4 3 2 1 0
SPIN 12097272

Preface

This volume reflects the proceedings of an International Workshop held in Paris in December 2006. The aim of the meeting was to undertake a high level of scholarly exchange between experts in cell biology, oncology and endocrinology interested in cell cycle control. The topics covered ranged from fundamental studies of DNA replication, chromosomal and nuclear function through growth factor control of endocrine tumor initiation and progression. The high quality chapters reflect the depth and breadth of the workshop lectures and discussions. Hopefully, the basic and translational insights gained from this book will be of interest to those studying the biology of endocrine tumors, as well as those deriving novel therapeutic approaches for these benign and malignant disorders.

I would like to acknowledge my co-program chairs, Drs. Henri Rochefort, Philippe Chanson and especially the wonderful Yves Christen for their superb organizational planning. We are indebted to the staff of Fondation Ipsen, including Jacqueline Mervaillie and Astrid de Gerard for their professionalism and skills in arranging the successful proceedings.

Shlomo Melmed

Table of Contents

List of Contributors

Anderson, Luke R.
Cancer Research Program, Garvan Institute of Medical Research, 384 Victoria St, Darlinghurst NSW 2010, Australia

Ben-Shlomo, Anat
Cedars-Sinai Medical Center, 8700 Beverly Blvd., Room 2015, Los Angeles, CA 90048, USA

Butt, Alison J.
Cancer Research Program, Garvan Institute of Medical Research, St Vincent's Hospital, 384 Victoria St, Darlinghurst NSW 2010, Australia

Chesnokova, Vera
Cedars-Sinai Medical Center, 8700 Beverly Blvd., Room 2015, Los Angeles, CA 90048, USA

Clarke, Robert B.
Breast Biology Group, Division of Cancer Studies, Faculty of Medicine and Human Sciences, University of Manchester, Paterson Institute for Cancer Research, Wilmslow Road, M20 4BX, Manchester, UK

Clemmons, David R.
CB# 7170, 8024 Burnett-Womack, Division of Endocrinology, University of North Carolina at Chapel Hill, Chapel Hill, NC 27599-7170, USA

Clurman, Bruce E.
Divisions of Clinical Research and Human Biology, Dana-Farber Cancer Institute and Department of Pathology, Harvard Medical School, Boston, Massachusetts 02115, USA

Comstock, Clay E.S.
Department of Cell Biology, University of Cincinnati College of Medicine, 3125 Eden Ave., Vontz Center, Cincinnati OH 45267 0521, USA

Geng, Yan
Department of Cancer Biology, Dana-Farber Cancer Institute and Department of Pathology, Harvard Medical School, Boston, Massachusetts 02115, USA

Harrison, Hannah
Breast Biology Group, Division of Cancer Studies, Faculty of Medicine and Human Sciences, University of Manchester, Paterson Institute for Cancer Research, Wilmslow Road, M20 4BX, UK

Inman, Claire K.
Cancer Research Program, Garvan Institute of Medical Research, St Vincent's Hospital, 384 Victoria St, Darlinghurst NSW 2010, Australia

Kaldis, Philipp
Mouse Cancer Genetics Program, National Cancer Institute, Frederick, Maryland 21702, USA

Knudsen, Karen E
Department of Cell Biology, University of Cincinnati College of Medicine, 3125 Eden Ave., Vontz Center, Cincinnati OH 45267 0521, USA

Lamb, Rebecca
Breast Biology Group, Division of Cancer Studies, Faculty of Medicine and Human Sciences, University of Manchester, Paterson Institute for Cancer Research, Wilmslow Road, M20 4BX, UK

Lee, Youngmi
Department of Cancer Biology, Dana-Farber Cancer Institute and Department of Pathology, Harvard Medical School, Boston, Massachusetts 02115, USA

Li, Jonathan J.
Hormonal Oncogengesis Laboratory, Department of Pharmacology, Toxicology and Therapeutics, The University of Kansas Medical Center, 3901 Rainbow Boulevard, Kansas City, KS, 66160, USA

Li, Sara Antonia
Hormonal Oncogengesis Laboratory, Department of Pharmacology, Toxicology and Therapeutics, The University of Kansas Medical Center, 3901 Rainbow Boulevard, Kansas City, KS, 66160, USA

McNeil, Catriona M.
Cancer Research Program, Garvan Institute of Medical Research, St Vincent's Hospital, 384 Victoria St, Darlinghurst NSW 2010, Australia

Méchali, Marcel
Institute of Human Genetics, CNRS, 141 Rue de la Cardonille, 34980, Montpellier, France

Melmed, Shlomo
Cedars-Sinai Medical Center, 8700 Beverly Blvd., Room 2015, Los Angeles, CA 90048, USA 423-0119

Musgrove, Elizabeth A.
Cancer Research Program, Garvan Institute of Medical Research, St Vincent's Hospital, 384 Victoria St, Darlinghurst NSW 2010, Australia

Olshavsky, Nicholas A.
Department of Cell Biology, University of Cincinnati College of Medicine, 3125 Eden Ave., Vontz Center, Cincinnati OH 45267 0521, USA

Pestell, Richard G.
Kimmel Cancer Center, Department of Cancer Biology, Thomas Jefferson University, Philadelphia, PA, USA

Pines, Jonathon
Wellcome/Cancer Research UK, Gurdon Institute, Tennis Court Road, Cambridge CB2 1QN, UK

Popov, Vladimir M.
Kimmel Cancer Center, Department of Cancer Biology, Thomas Jefferson University, Philadelphia, PA, USA

Powell, Michael J.
Kimmel Cancer Center, Department of Cancer Biology, Thomas Jefferson University, Philadelphia, PA, USA

Roberts, James M.
Howard Hughes Medical Institute, Division of Basic Sciences, Fred Hutchinson Cancer Research Center, Seattle, Washington 98109, USA

Rochefort, Henri
INSERM Hormones and Cancers Units (U148, U540) and University Montpellier 1; Centre de recherche en cancérologie. CRLC Val d'Aurelle, 34298 Montpellier Cedex 5, France

Sergio, C. Marcelo
Cancer Research Program, Garvan Institute of Medical Research, St Vincent's Hospital, 384 Victoria St, Darlinghurst NSW 2010, Australia

Sharma, Ankur
Department of Cell Biology, University of Cincinnati College of Medicine, 3125 Eden Ave., Vontz Center, Cincinnati OH 45267 0521, USA

Sicinski, Piotr
Department of Cancer Biology, Dana-Farber Cancer Institute and Department of Pathology, Harvard Medical School, Boston, Massachusetts 02115, USA

Slingerland, Joyce
Braman Family Breast Cancer Institute, Sylvester Comprehensive Cancer Center, University of Miami School of Medicine, 1475 N.W. 12th Avenue (D8-4), Miami, FL 33136 USA

Sutherland, Robert L.
Cancer Research Program, Garvan Institute of Medical Research, St Vincent's Hospital, 384 Victoria St, Darlinghurst NSW 2010, Australia

Swanger, Jherek
Divisions of Clinical Research and Human Biology, Dana-Farber Cancer Institute and Department of Pathology, Harvard Medical School, Boston, Massachusetts 02115, USA

Wang, Chenguang
Kimmel Cancer Center, Department of Cancer Biology, Thomas Jefferson University, Philadelphia, PA, USA

Welcker, Markus
Divisions of Clinical Research and Human Biology, Dana-Farber Cancer Institute and Department of Pathology, Harvard Medical School, Boston, Massachusetts 02115, USA

Yu, Run
Cedars-Sinai Medical Center, 8700 Beverly Blvd., Room 2015, Los Angeles, CA 90048, USA

Zagozdzon, Agnieszka
Department of Cancer Biology, Dana-Farber Cancer Institute and Department of Pathology, Harvard Medical School, Boston, Massachusetts 02115, USA

DNA Replication Origins, Development, and Cancer

Marcel Méchali[1]

Introduction

DNA replication is at the heart of living organisms. The main purpose of any organism, even the simplest, is the duplication of the genetic information and its transmission to the offspring. In simple organisms, such as the bacteria *Echerichia coli*, a single DNA replication origin is used, from which replication forks migrate in opposite directions to allow the duplication of the entire genome. In complex eukaryotes, such as multicellular organisms, multiple DNA replication origins are used to replicate chromosomes, which are highly regulated during S phase. Here, I will review how the positions of these origins are regulated and how DNA replication origins might be regulatory elements that also control development and cell identity. I will also consider how DNA replication origins are regulated within each cell cycle in order to avoid re-replication, an event that might lead to genetic damage and neoplasic evolution.

Initiation of DNA Replication

Forty-four years ago, the replicon model was proposed (Jacob et al. 1963) to explain the initiation of DNA replication in bacteria. The model proposed that a specific sequence in the DNA of *E. coli* determines the origin of DNA replication and that specific protein(s) that recognize this sequence allow the opening of the DNA double helix at this position. Since then, the sequence-specific nature of the initiation of DNA replication has been confirmed in the prokaryotic world. This concept also proved to be correct for the replication of eukaryotic small DNA virus, and the model was further extended to the replication of the unicellular organism, *Saccharomyces cerevisiae* (Aladjem et al. 2006).

In multicellular organisms, the notion of an origin of DNA replication remains rather elusive. In human cells, around 50 000 DNA replication origins are used during each S-phase. These origins are spaced at around 100 Kbp intervals. However, in contrast to the yeast *S. cerevisiae*, in which a consensus sequence is common to all its origins, no common sequence has yet been found at metazoan DNA replication origins. This is the main paradox in the field because, although replication origins do not share common specific sequence elements, they appear to be at specific positions along the chromosomes. One hypothesis proposed to explain this paradox is that the evolution from prokaryotic to complex metazoan genomes may have introduced additional

[1] Institute of Human Genetics, CNRS, 141 Rue de la Cardonille, 34980, Montpellier, France
e-mail: marcel.mechali@igh.cnrs.fr

Melmed et al.
Hormonal Control of Cell Cycle
© Springer-Verlag Berlin Heidelberg 2008

functions of replication origins that allow the coupling of DNA replication to gene expression, particularly when cell growth has to be coordinated to cell differentiation (Méchali 2001).

One difficulty in resolving this paradox has a technical cause. Although autoradiographic examinations of DNA fibers have rapidly established the mean size of a replication unit and the average number of DNA replication origins (Huberman and Riggs 1968), origin identification at the sequence level is still a complex task. DNA replication origins are activated asynchronously with a specific temporal timing, at least for some of them, and there is as yet no specific functional assay equivalent to the assay developed in yeast. In yeast, autonomously replicating sequences (ARS) confer the ability to recombinant plasmids to replicate in an extrachromosomal manner after transfection. Such an assay does not work in human cells, where any sequence inserted into a plasmid molecule can replicate (albeit poorly), providing it has a minimum size (Heinzel et al. 1991; Krysan and Calos 1991, 1993). Several methods have been developed to map DNA replication origins at chosen DNA domains. The most widely used now is based on the purification of short (1–2 Kbp) RNA primed nascent DNA molecules that are synthesized at DNA replication origins, followed by their quantification by real time PCR using several primers along the origin of interest. This method is accurate but requires large amounts of cells. Indeed, nascent strands at a given origin will only be detected within a very short period of time, as DNA replication forks migrate at 3 Kb/min. Moreover, it is practically impossible to define within minutes the precise timing of activation of a given DNA replication origin during the six- to eight-hour S phase period. It is therefore not surprising that less than 0.1% of DNA replication origins have been mapped in metazoans. The rapid improvements in the use of DNA microarrays or DNA combing will certainly result in a large improvement in our knowledge of DNA replication origins in the near future (Schubeler et al. 2002; MacAlpine et al. 2004; Jeon et al. 2005).

DNA Replication Origins in Somatic Cells and Their Characteristics

Variable Size

In metazoans, site-specific initiation of DNA replication has been validated for 20 to 30 origins (Aladjem et al. 2006). Often, the mapping is not precise and has to be confirmed. At least three DNA replication origins have been carefully examined in different conditions: the dihydrofolate reductase (DHFR) origin, the lamin B origin, and the c-myc origin. The laboratory of Joyce Hamlin first designed a cell line in which the DHFR gene was amplified 1 000 times, and the positions of the DNA replication origins in the locus were determined by several different methods and different laboratories, sometimes with controversial results (DePamphilis 2003). Overall, the results suggest that there is a large potential initiation zone in this locus, where at least two preferential origins are used.

The DNA replication origin found downstream of the lamin B2 gene exhibits opposite properties. This origin was found to be precisely positioned at the nucleotide level (Abdurashidova et al. 2000), and it is also active at an ectopic position (Paixao et al. 2004).

The c-myc DNA replication origin has also been extensively analyzed. It is located in a 2 Kbp region upstream of the gene (Liu et al. 2003; Ghosh et al. 2006). This origin has been found in the same region in different species (human, mouse, Xenopus).

Sequence Specificity

One of the main characteristics of the DNA replication origins identified in metazoans is the lack of a clear consensus DNA sequence, suggesting that elements other than sequence recognition contribute to the definition of a replication origin in complex eukaryotes. Another characteristic is the lack of a defined size of the DNA replication origin. The DHFR origin and the lamin B origin have clearly opposed properties in this respect.

By analogy to the sequence-specific DNA replication origins in *S. cerevisiae*, it has been suggested that the metazoan origin might have a modular structure, although each module contributing to the origin activity is not conserved among origins. The c-myc origin as well as the β globin origin exhibit adjacent modules that may act in synergy to achieve efficient initiation (Liu et al. 2003; Wang et al. 2006). Among such modules, the property often identified is A-T bases richness, mostly asymmetric (Abdurashidova et al. 2000; DePamphilis 2003; Altman and Fanning 2004; Wang et al. 2006). This is also a property of the consensus ARS sequence in *S. cerevisiae*, although it is not sufficient to confer efficient autonomous DNA replication in yeast. An AT-rich sequence may obviously help to open the DNA double helix at the origin. They could also be recognized by specific factors such as AT hook domain proteins. In *S. pombe*, where no consensus sequence can be defined for replication origins, ORC4, a subunit of the ORC complex that binds origins of DNA replication, contains nine AT hook domains (Chuang and Kelly 1999). In metazoans, ORC subunits do not contain such domains but also have the ability to preferentially bind AT-rich elements (Kong et al. 2003; Vashee et al. 2003).

Epigenetic Regulation

Epigenetic controls have often been suspected of contributing to DNA replication origin specification (Méchali 2001). Among them, transcription is the most obvious candidate. Transcription factors or chromatin remodelling proteins involved in transcription may help to recruit DNA replication initiation proteins to origins. However, the function of proteins from the transcription apparatus might not always define a replication origin. By displacing histones from DNA, the process of transcription may contribute to the formation of a chromatin domain in which the DNA double helix will be easier to open. This possible synergy between transcription and replication factors has been well documented in the replication of viral DNAs, such as SV40, polyoma, or BPV (Kohzaki and Murakami 2005). In yeast, binding sites for transcription factors have also been found close to ARS sequences. The B3 element, found in ARS1, contains a site for Abf1, which regulates the expression of a large number of genes (Li and Rosenfeld 2004). MCM1, a transcription factor essential in yeast, is also involved in the regulation of the activity of ARS elements (Chang et al. 2003).

In metazoans, transcription factors of the Hox family were also found at the human lamin B2 replication origin (de Stanchina et al. 2000). The possible involvement of transcription factors in the initiation of DNA replication has been well documented in Drosophila, at the chorion gene amplified locus, which contains a specific origin of replication. A control element (ACE3) binds several transcription factors, including E2F1, Myb, and Mip120 (Beall et al. 2002). However, this DNA replication origin is involved in DNA amplification of a specific locus during development, a phenomenon that does not occur during normal cell proliferation. The interconnection between transcription and replication remains an attractive hypothesis in DNA replication origin specification. Nevertheless, clear evidence for a global role of transcription factors in initiation of DNA replication remains difficult to obtain. One caveat in this issue is the nature of the DNA replication origins examined, which are often close to promoter regions and therefore where several sites for transcription activators and replication activators may overlap without being functionally linked.

Modification of chromatin structure is a second possible epigenetic contribution to initiation of DNA replication. The chromosomal environment has been identified as a possible regulator of DNA replication origins in yeast. ARS elements active in plasmids are not always active in their chromosomal context (Fangman and Brewer 1992). In addition, ARS elements can be active in their normal chromosomal context but not when displaced to another chromosomal region.

Acetylation of histones is a process required for transcriptional activation. As for DNA replication, this regulation was supported with results obtained in Drosophila and Xenopus. In Xenopus, the formation of a transcription complex on a recombinant DNA molecule induces the specification of a DNA replication origin, and histones appear to be specifically acetylated at the origin (Danis et al. 2004). Similarly, inhibition of histone acetylation strongly inhibits chromosomal DNA replication in Xenopus eggs (Lemaitre et al. 2005). Histones are also hyperacetylated at the chorion gene DNA replication origin (Aggarwal and Calvi 2004). However, this phenomenon might not be universal. In the chicken β globin locus, DNA replication origins are found in both hypoacetylated and acetylated regions (Dazy et al. 2006). In undifferentiating P19 teratocarcinoma cells, the main HoxB DNA replication origin appears hypoacetylated (Gregoire et al. 2006, Cayrou and Méchali, unpublished).

DNA Replication Origins and the Regulation of Development

In 1972, the discovery that SV40, a virus infecting animal cells, had a single precise and sequence-specific DNA replication origin extended the replicon model to the eukaryotic world (Tegtmeyer 1972). However, soon thereafter, Harland and Laskey (1980) showed that any DNA fragment from the SV40 genome could replicate when injected into Xenopus eggs. This result was quite unexpected and led to some controversy in the field. However, it was soon confirmed and further shown that the efficiency of DNA replication in the system did not rely on the sequence but on the size of the injected DNA molecules (Méchali et al. 1983). This lack of sequence specificity was further confirmed in the endogenous replicating genome during early development in both Xenopus and Drosophila (Hyrien and Méchali 1993; Sasaki et al. 1999). Importantly, in both cases, the embryonic genome is not transcribed during this period, leading to the possibility that, at least for some genes, DNA replication origins may become site-specific only when chromosomal domains are organized for transcription. This conjecture was confirmed to be the case when it was demonstrated that DNA replication origins become site-specific in the rDNA gene cluster only after the mid blastula transition, when transcription resumes in the embryo. This correlation between transcription and site-specific initiation of DNA replication was finally experimentally reproduced by injection of recombinant DNA molecules into Xenopus eggs. DNA replication occurs in such conditions but with no site specificity (Harland and Laskey 1980; Méchali and Harland 1982; Méchali et al. 1983), unless the plasmid is made competent for transcription by the assembly of a transcription complex on a promoter cloned into the plasmid molecule. In this case, a major DNA replication origin is detected at or close to the promoter (Danis et al. 2004). Therefore, even in the context of early development, a site-specific DNA replication origin can be fixed, provided that the chromatin domain is programmed for transcription. Interestingly, transcription itself is not necessary. Results obtained in *Sciara coprophila* also show that the OriII/9A origin appears to be more defined in cells determined for transcription but not yet engaged in active transcription (Lunyak et al. 2002).

This role of the establishment of transcription programs in the regulation of the DNA replication origin pattern explained observations made in the 1970s showing that, in Drosophila development, the spacing between DNA replication origins varied between early development and adult tissues (Blumenthal et al. 1974). A short spacing of DNA replication origins is required in Xenopus early development, as cell division is controlled by a cell cycle clock activated every 30 min (Hara et al. 1980). The S phase occurs in less than 15 min, so it is crucial that DNA replication origins are closely and evenly spaced to avoid regions of unreplicated DNA at the time of cell division.

Recently, a role for chromosome structure in the positioning of DNA replication origins has been demonstrated (Lemaitre et al. 2005) that may explain why animal cloning is so inefficient (less than 8%) in most animal species. When sperm nuclei are incubated in Xenopus egg extracts, they replicate as efficiently as after fertilization, during the early embryonic development. In contrast, somatic cell nuclei incubated in the same egg extract replicate rather poorly. One explanation for this low efficiency is that somatic cell chromosomes have some imprinted structure characteristic of DNA replication in somatic tissues and that this structure is not adapted to early embryonic development. However, if somatic cell nuclei are incubated in egg extracts blocked in mitosis, they acquire the structure of embryonic mitotic chromosomes and can then replicate their DNA as efficiently as sperm nuclei. The spacing of DNA replication origins, which is 120 Kbp in somatic cell nuclei, is reduced to 20 Kbp, as for early embryonic chromosomes, allowing rapid and efficient DNA replication. When the size of chromatin loops is measured, it is found to be the same as the size of the replication unit. Interestingly, this mitotic remodelling of the eukaryotic replicon is dependent on DNA topoisomerase II, an enzyme found at the nuclear matrix. Altogether, these data suggest that incubation of somatic cell nuclei in mitotic egg extract permits the reprogramming of the global chromosome structure in such a way that the pattern of replication origins in now adapted to the developmental program.

Changes in the DNA replication origin pattern have also been observed in other organisms, such as the slime mold *Physarum polycephalum* (Maric et al. 2003). Here, for two genes – one expressed in the ameba stage (profilin A) and the other in the plasmodia (profilin B) – the DNA replication origin is present only at the gene being expressed in the particular cell type. In mouse pluripotent P19 cells, when the Hox B gene is not transcribed, more than one origins is present over the domain, but when cells are engaged in differentiation, a single DNA replication origin is used (Gregoire et al. 2006). Similarly, in murine non-B cells, a DNA replication origin is localized downstream of the IgH locus, whereas in pro B and pre B cell lines, multiple initiation sites are used (Zhou et al. 2002). Therefore, studies during both in vivo development and in vitro cell differentiation suggest that DNA replication origins are not rigidly fixed in totipotent or pluripotent cells.

DNA Replication Origins and Cancer

The Replication Licensing Reaction

Initiation of DNA replication can be regulated both at the level of the specificity of the localization of replication origins and by the frequency at which the origins are used during a cell cycle. The relatively flexible nature of metazoan replication origins suggests that the positioning of origins may be altered in cancer (although this has not yet been demonstrated). A normal eukaryotic cell should tightly control the frequency of DNA replication origin firing in such a way that none of the 50 000 DNA replication origins is activated more that once during each cell cycle. The occurrence of such an event would result in the amplification of the corresponding chromatin domain, a phenomenon that often occurs in cancer cells.

Factors Involved in Replication Licensing and Their Importance in Cancer

Several proteins have been discovered in recent years that bind to DNA replication origins and prepare the genome for S phase (Sivaprasad et al. 2006). These proteins assemble the pre-replicative complex onto origins during G1. The first protein known to recognize the DNA replication origin is the ORC complex. Then, two factors are recruited, Cdc6 and Cdt1, which themselves will recruit the DNA helicase, the MCM2-7 complex, to open the double helix at the origin. Once the DNA helicase is loaded, the DNA is "licensed" for DNA replication, and the DNA helicase can unwind the origin, allowing the polymerase machinery and its cofactors to enter to synthesize the complementary DNA strand.

Two factors are essential to control the initiation of DNA synthesis at the origin and to prevent more than one event of initiation at each origin: cdt1 and geminin, which act as a switch at the origins. Cdt1 is loaded onto DNA replication origins in G1, after the ORC complex is assembled (Maiorano et al. 2000), and is required to load MCM2-7. Once the MCMs are loaded, nuclei are competent for DNA synthesis and Cdt1 is not required anymore. Soon after initiation of DNA synthesis has started, cdt1 is removed from chromatin. Cdt1 is negatively controlled by geminin, which does not permit reinitiation to occur at already licensed origins (McGarry and Kirschner 1998; Wohlschlegel et al. 2000; Tada et al. 2001). Recent data have shown that Cdt1 and geminin form a complex that contains the ability to both activate and inhibit DNA replication origins. A Cdt1-geminin complex is rapidly loaded onto chromatin in G1, allowing initiation of S phase, but as soon origins are activated, other geminin molecules are added to the complex and prevent relicensing of origins. Switching the origin on and off through the

Cdt1-geminin complex is therefore an important regulation that prevents illegitimate reinitiation of DNA replication during the cell cycle. If the stochiometric ratio of Cdt1 to geminin is important to define the on and off stage, then one would expect that disrupting this ratio might lead to rereplication phenotypes. This rereplication was indeed the case, in Drosophila, Xenopus, and human cells. In Drosophila, geminin mutants exhibited rereplication (Quinn et al. 2001; Mihaylov et al. 2002). In Xenopus, a high level of Cdt1 also promoted rereplication (Walter and Newport 1997, 2000; Maiorano et al. 2005; Lutzmann et al. 2006). In human cells, siRNA silencing of geminin led to rereplication (Melixetian et al. 2004; Zhu et al. 2004). The importance of Cdt1 and geminin in the fine control of origin firing led to the question of whether Cdt1 was an oncogene and geminin a tumor suppressor. It was first reported that murine NIH3T3 cells can promote tumor formation in mice when Cdt1 is overexpressed (Arentson et al. 2002), but, recently, no phenotype was detected with Rat 1 cells overexpressing Cdt1 (Tatsumi et al. 2006). It is possible that genomic instability induced by high levels of Cdt1 has some impact in tumor progression but that the consequences vary according to the properties of the cell line. As for geminin, its suggested function as a tumor suppressor is in apparent contradiction with the observation that geminin is overexpressed in tumor cells (Wohlschlegel et al. 2002; Gonzalez et al. 2004; Wharton et al. 2004). However, geminin also has important functions in cell differentiation (Del Bene et al. 2004; Luo et al. 2004), and its overexpression may deregulate the balance between cell proliferation and cell differentiation.

So far, no specific mutations in proteins involved in DNA replication origin licensing have been described in tumors. High levels of MCM proteins have been shown to be a good marker for cancer progression in several cases (Gonzalez et al. 2005; Coleman et al. 2006). It is possible that it is the stochiometry of the different proteins forming the pre-replication complex that is important for regulating DNA replication. If this is the case, the use of DNA replication initiation markers for cancer prognostics might rely on the relative expression of these markers at the protein level for each kind of tumor.

References

Abdurashidova G, Deganuto M, Klima R, Riva S, Biamonti G, Giacca M, Falaschi A (2000) Start sites of bidirectional DNA synthesis at the human lamin B2 origin. Science 287:2023–2026

Aggarwal BD, Calvi BR (2004) Chromatin regulates origin activity in Drosophila follicle cells. Nature 430:372–376

Aladjem MI, Falaschi A, Kowalski D (2006) Eukaryotic DNA replication origins. In: DNA replication and human disease. (ML Depamphilis ed) CSHL Press, pp. 31–62

Altman AL, Fanning E (2004) Defined sequence modules and an architectural element cooperate to promote initiation at an ectopic mammalian chromosomal replication origin. Mol Cell Biol 24:4138–4150

Arentson E, Faloon P, Seo J, Moon E, Studts JM, Fremont DH, Choi K (2002) Oncogenic potential of the DNA replication licensing protein CDT1. Oncogene 21:1150–1158

Beall EL, Manak JR, Zhou S, Bell M, Lipsick JS, Botchan MR (2002) Role for a Drosophila Myb-containing protein complex in site-specific DNA replication. Nature 420:833–837

Blumenthal AB, Kriegstein HJ, Hogness DS (1974) The units of DNA replication in Drosophila melanogaster chromosomes. Cold Spring Harb Symp Quant Biol 38:205–223

Chang VK, Fitch MJ, Donato JJ, Christensen TW, Merchant AM, Tye BK (2003) Mcm1 binds replication origins. J Biol Chem 278:6093–6100

Chuang RY, Kelly TJ (1999) The fission yeast homologue of Orc4p binds to replication origin DNA via multiple AT-hooks. Proc Natl Acad Sci USA 96:2656–2661

Coleman N, Mills AD, Laskey R (2006) Cancer diagnosis and DNA replication. In: DNA replication and human disease. CSHL Press (ML Depamphilis ed), pp. 501–518

Danis E, Brodolin K, Menut S, Maiorano D, Girard-Reydet C, Méchali M (2004) Specification of a DNA replication origin by a transcription complex. Nature Cell Biol 6:721–730

Dazy S, Gandrillon O, Hyrien O, Prioleau MN (2006) Broadening of DNA replication origin usage during metazoan cell differentiation. EMBO Rep 7:806–811

de Stanchina E, Gabellini D, Norio P, Giacca M, Peverali FA, Riva S, Falaschi A, Biamonti G (2000) Selection of homeotic proteins for binding to a human DNA replication origin. J Mol Biol 299:667–680

Del Bene F, Tessmar-Raible K, Wittbrodt J (2004) Direct interaction of geminin and Six3 in eye development. Nature 427:745–749

DePamphilis ML (2003) Eukaryotic DNA replication origins:reconciling disparate data. Cell 114:274–275

Fangman WL, Brewer BJ (1992) A question of time:replication origins of eukaryotic chromosomes. Cell 71:363–366

Ghosh M, Kemp M, Liu G, Ritzi M, Schepers A, Leffak M (2006) Differential binding of replication proteins across the human c-myc replicator. Mol Cell Biol 26:5270–5283

Gonzalez MA, Tachibana KE, Chin SF, Callagy G, Madine MA, Vowler SL, Pinder SE, Laskey RA, Coleman N (2004) Geminin predicts adverse clinical outcome in breast cancer by reflecting cell-cycle progression. J Pathol 204:121–130

Gonzalez MA, Tachibana KE, Laskey RA, Coleman N (2005) Control of DNA replication and its potential clinical exploitation. Nature Rev Cancer 5:135–141

Gregoire D, Brodolin K, Méchali M (2006) HoxB domain induction silences DNA replication origins in the locus and specifies a single origin at its boundary. EMBO Rep 7:812–816

Hara K, Tydeman P, Kirschner M (1980) A cytoplasmic clock with the same period as the division cycle in Xenopus eggs. Proc Natl Acad Sci USA 77:462–466

Harland RM, Laskey RA (1980) Regulated replication of DNA microinjected into eggs of Xenopus laevis. Cell 21:761–771

Heinzel SS, Krysan PJ, Tran CT, Calos MP (1991) Autonomous replication in human cells is affected by the size and the source of DNA. Mol Cell Biol 11:2263–2272

Huberman JA, Riggs AD (1968) On the mechanism of DNA replication in mammalian chromosomes. J Mol Biol 32:327–341

Hyrien O, Méchali M (1993) Chromosomal replication initiates and terminates at random sequences but at regular intervals in the ribosomal DNA of Xenopus early embryos. EMBO J 12:4511–4520

Jacob F, Brenner J, Cuzin F (1963) On the regulation of DNA replication in bacteria. Cold Spring Harbor Symp Quant Biol 28:329–348

Jeon Y, Bekiranov S, Karnani N, Kapranov P, Ghosh S, MacAlpine D, Lee C, Hwang DS, Gingeras TR, Dutta A (2005) Temporal profile of replication of human chromosomes. Proc Natl Acad Sci USA 102:6419–6424

Kohzaki H, Murakami Y (2005) Transcription factors and DNA replication origin selection. Bioessays 27:1107–1116

Kong D, Coleman TR, DePamphilis ML (2003) Xenopus origin recognition complex (ORC) initiates DNA replication preferentially at sequences targeted by Schizosaccharomyces pombe ORC. Embo J 22:3441–3450

Krysan PJ, Calos MP (1991) Replication initiates at multiple locations on an autonomously replicating plasmid in human cells. Mol Cell Biol 11:1464–1472

Krysan PJ, Calos MP (1993) Epstein-Barr virus-based vectors that replicate in rodent cells. Gene 136:137–143

Lemaitre JM, Danis E, Pasero P, Vassetzky Y, Méchali M (2005) Mitotic remodeling of the replicon and chromosome structure. Cell 123:1–15

Li X, Rosenfeld MG (2004.)Transcription: origins of licensing control. Nature 427:687–678

Liu G, Malott M, Leffak M (2003) Multiple functional elements comprise a Mammalian chromosomal replicator. Mol Cell Biol 23:1832–1842

Lunyak VV, Ezrokhi M, Smith HS, Gerbi SA (2002) Developmental Changes in the Sciara II/9A Initiation Zone for DNA Replication. Mol Cell Biol 22:8426–8437

Luo L, Yang X, Takihara Y, Knoetgen H, Kessel M (2004) The cell-cycle regulator geminin inhibits Hox function through direct and polycomb-mediated interactions. Nature 427:749–753

Lutzmann M, Maiorano D, Méchali M (2006) A Cdt1-geminin complex licenses chromatin for DNA replication and prevents rereplication during S phase in Xenopus. Embo J 25:5764–5774

MacAlpine DM, Rodriguez HK, Bell SP (2004) Coordination of replication and transcription along a Drosophila chromosome. Genes Dev 18:3094–3105

Maiorano D, Moreau J, Méchali M (2000) XCDT1 is required for the assembly of pre-replicative complexes in Xenopus laevis. Nature 404:622–625

Maiorano D, Krasinska L, Lutzmann M, Méchali M (2005) Recombinant Cdt1 induces rereplication of G2 nuclei in Xenopus egg extracts. Curr Biol 15:146–153

Maric C, Benard M, Pierron G (2003) Developmentally regulated usage of Physarum DNA replication origins. EMBO Rep 4:474–478

McGarry TJ, Kirschner MW (1998) Geminin, an inhibitor of DNA replication, is degraded during mitosis. Cell 93:1043–1053

Méchali M (2001) DNA replication origins: from sequence specificity to epigenetics. Nature Rev Genet 2:640–645

Méchali M, Harland RM (1982) DNA synthesis in a cell-free system from Xenopus eggs:priming and elongation on single-stranded DNA in vitro. Cell 30:93–101

Méchali M, Méchali F, Laskey RA (1983) Tumor promoter TPA increases initiation of replication on DNA injected into xenopus eggs. Cell 35:63–69

Melixetian M, Ballabeni A, Masiero L, Gasparini P, Zamponi R, Bartek J, Lukas J, Helin K (2004) Loss of Geminin induces rereplication in the presence of functional p53. J Cell Biol 165:473–482

Mihaylov IS, Kondo T, Jones L, Ryzhikov S, Tanaka J, Zheng J, Higa LA, Minamino N, Cooley L, Zhang HL (2002) Control of DNA replication and chromosome ploidy by geminin and cyclin A. Mol Cell Biol 22:1868–1880

Paixao S, Colaluca IN, Cubells M, Peverali FA, Destro A, Giadrossi S, Giacca M, Falaschi A, Riva S, Biamonti GS (2004) Modular structure of the human lamin B2 replicator. Mol Cell Biol 24:2958–2967

Quinn LM, Herr A, McGarry TJ, Richardson H (2001) The Drosophila Geminin homolog:roles for Geminin in limiting DNA replication, in anaphase and in neurogenesis. Genes Dev 15:2741–2754

Sasaki T, Sawado T, Yamaguchi M, Shinomiya T (1999) Specification of regions of DNA replication initiation during embryogenesis in the 65-kilobase DNApolalpha-dE2F locus of Drosophila melanogaster. Mol Cell Biol 19:547–555

Schubeler D, Scalzo D, Kooperberg C, van Steensel B, Delrow J, Groudine M (2002) Genome-wide DNA replication profile for Drosophila melanogaster: a link between transcription and replication timing. Nature Genet 32:438–442

Sivaprasad U, Dutta A, Bell SP (2006) Assembly of pre-replication complexes. In: DNA replication and human disease. CSHL Press (ML Depamphilis ed), pp. 63–88

Tada S, Li A, Maiorano D, Méchali M, Blow JJ (2001) Repression of origin assembly in metaphase depends on inhibition of RLF- B/Cdt1 by geminin. Nature Cell Biol 3:107–113

Tatsumi Y, Sugimoto N, Yugawa T, Narisawa-Saito M, Kiyono T, Fujita M (2006) Deregulation of Cdt1 induces chromosomal damage without rereplication and leads to chromosomal instability. J Cell Sci 119:3128–3140

Tegtmeyer P (1972) Simian virus 40 deoxyribonucleic acid synthesis: the viral replicon. J Virol 10:591–598

Vashee S, Cvetic C, Lu W, Simancek P, Kelly TJ, Walter JC (2003) Sequence-independent DNA binding and replication initiation by the human origin recognition complex. Genes Dev 17:1894–1908

Walter J, Newport JW (1997) Regulation of replicon size in Xenopus egg extracts. Science 275:993–995

Walter J, Newport J (2000) Initiation of eukaryotic DNA replication:origin unwinding and sequential chromatin association of Cdc45, RPA, and DNA polymerase alpha. Mol Cell 5:617–627

Wang L, Lin CM, Lopreiato JO, Aladjem MI (2006) Cooperative sequence modules determine replication initiation sites at the human beta-globin locus. Human Mol Genet 15:2613–2622

Wharton SB, Hibberd S, Eward KL, Crimmins D, Jellinek DA, Levy D, Stoeber K, Williams GH (2004) DNA replication licensing and cell cycle kinetics of oligodendroglial tumours. Br J Cancer 91:262–269

Wohlschlegel JA, Dwyer BT, Dhar SK, Cvetic C, Walter JC, Dutta A (2000) Inhibition of eukaryotic DNA replication by geminin binding to Cdt1. Science 290:2309–2312

Wohlschlegel JA, Kutok JL, Weng AP, Dutta A (2002) Expression of geminin as a marker of cell proliferation in normal tissues and malignancies. Am J Pathol 161:267–273

Zhou J, Ermakova OV, Riblet R, Birshtein BK, Schildkraut CL (2002) Replication and subnuclear location dynamics of the immunoglobulin heavy-chain locus in B-lineage cells. Mol Cell Biol 22:4876–4889

Zhu W, Chen Y, Dutta A (2004) Rereplication by depletion of geminin is seen regardless of p53 status and activates a G2/M checkpoint. Mol Cell Biol 24:7140–7150

Getting In and Out of Mitosis

Jonathon Pines[1]

Summary

There are two major problems for the cell to solve in mitosis: how to ensure that each daughter cell receives an equal and identical complement of the genome, and how to prevent cell separation before chromosome segregation. Both these problems are solved by controlling when two specific proteins are destroyed: securin, an inhibitor of chromosome segregation, and cyclin B, which inhibits cell separation (cytokinesis). It has recently become clear that a number of other proteins are degraded at specific points in mitosis. This review will focus on how specific proteins are selected for proteolysis at defined points in mitosis and how this selection contributes to the proper coordination of chromosome segregation and cytokinesis.

Introduction

Mitosis is conventionally divided up into discrete stages according to the morphology of the cell (but see Pines and Rieder 2001). In cells that undergo an open mitosis prophase ends with nuclear envelope breakdown and the subsequent stage, prometaphase, is defined by the search and capture behavior of microtubules as the kinetochores are attached to the spindle. Once all the kinetochores have correctly attached to the mitotic spindle the cell is defined as in metaphase, and chromosomes proceed to align on a "metaphase plate". By this definition metaphase can be a remarkably defined length of time that is likely to be set by how long it takes to degrade particular proteins (see below). Metaphase ends with the rapid and almost synchronous separation of all of the sister chromatids, which begin to segregate to opposite poles of the spindle (anaphase A), followed by elongation of the spindle itself (anaphase B). Once each set of sister chromatids has reached opposite spindle poles, they begin to decondense, the nuclear envelope reforms and the mitotic spindle disassembles (telophase). During anaphase and telophase the cell itself begins to divide (cytokinesis) to generate two genetically identical daughter cells, although in animal cells these do not complete separation until abscission that, in mammalian cell culture, can take place hours after cells re-enter interphase. A number of these events are coordinated by proteolysis.

Cells are driven into and through mitosis by the mitotic cyclin-dependent kinases (Cdk) working in concert with a number of other protein kinases, such as the Polo,

---

[1] Wellcome/Cancer Research, Gurdon Institute, Tennis Court Road, Cambridge CB2 1QN, UK

e-mail: jp103@cam.ac.uk

Melmed et al.
Hormonal Control of Cell Cycle
© Springer-Verlag Berlin Heidelberg 2008

Aurora and NIMA-families (reviewed in Nigg 2001). The kinases are often coordinated by recruitment to a substrate only after it has been phosphorylated by an upstream kinase. For example, polo kinases use their polo-box to bind to sites previously phosphorylated by mitotic CDKs (Elia et al. 2003). Moreover, the cyclins themselves have recently been shown to contain a phospho-peptide binding site in their conserved "cyclin fold" (Mimura et al. 2004). The mitotic kinases are antagonized by phosphatases, and it is the balance between these that controls a number of steps in mitosis. Although phosphorylation is a rapidly reversible event, inactivating a kinase or phosphatase by proteolysis can make phosphorylation or dephosphorylation effectively irreversible and can confer directionality.

The ability to select a specific protein for rapid proteolysis is conferred by the ubiquitin-proteasome system (UPS; reviewed in Hershko and Ciechanover 1998). Proteins degraded by the UPS are tagged with a multi-ubiquitin chain that is recognized by the proteasome cap. Proteasomes appear to be constitutively active throughout the cell cycle; therefore, substrate selection is primarily controlled by when and where proteins are ubiquitinated. Ubiquitin is transferred onto the ε-amino group of a lysine residue of a substrate by a ubiquitin conjugating enzyme (UBC) working in concert with a ubiquitin ligase. The ubiquitin can subsequently be removed by deubiquitinating enzymes (DUBs), a large number of which are encoded in the genome. Some DUBs are components of the proteasome cap, where they have a general role in "proofreading" substrate selection or recycling ubiquitin. It is highly likely that others will be found that are required for the proper regulation of mitosis. Although ubiquitination can target a protein to the proteasome, it can also perform other important roles, for example, in endocytosis and signal transduction. Thus, some proteins may be ubiquitinated in mitosis for purposes other than destruction.

Entry to mitosis can be regulated by proteolysis, but it is in mitosis itself that the UPS has its most defined cell cycle roles. In mitosis, most of the specificity in substrate selection is conferred by the ubiquitin ligase, of which the most prominent is a multi-subunit complex called the Anaphase Promoting Complex or Cyclosome (APC/C). The APC/C has the primary role in ensuring correct chromosome segregation and in coordinating mitosis with cytokinesis. Thus, the questions of how the APC/C is activated and how it recognizes its substrates are key to understanding how mitosis is regulated.

The Multi-faceted APC/C

The APC/C is composed of up to 13 different subunits in yeast and 11 subunits in animal cells (reviewed in Passmore 2004; Peters 2002). The catalytic subunits are APC11, a RING finger protein, APC2, a protein with homology to the cullin family, and Doc1, a subunit that is important for substrate recognition and/or extending the poly-ubiquitin chain on a substrate (Carroll et al. 2005; Carroll and Morgan 2002; Passmore et al. 2003). The function of the other subunits is unclear, but there is evidence that they may also be important in substrate recognition. A number of these subunits contain protein:protein interaction domains of the TPR family and are multiply phosphorylated in mitosis, which is required to activate the APC/C (Kraft et al. 2003; Rudner and Murray 2000) but could also alter substrate binding affinities. A number of the phosphorylation sites have been mapped and are mitotic cyclin-Cdk and Polo-like kinase sites, of which the cyclin-Cdk sites are the most important for activating the APC/C (Kraft et al. 2003; Rudner and Murray 2000).

In addition to its core components, the APC/C requires a member of the WD40 family for activity. Three different WD40 proteins can act with the APC/C: Cdc20, Cdh1 and Ama1 (these are the names of the proteins in budding yeast). These proteins have a conserved isoleucine-arginine (IR) dipeptide motif at their C-terminus that is required for them to bind to the APC/C, apparently to subunits with TPR motifs (Vodermaier et al. 2003). The WD40 proteins act at different times in the cell cycle and alter the range of substrates recognized by the APC/C (reviewed in Vodermaier 2001). Cdc20 (*fizzy* in *Drosophila*) acts in all cells; it is most important in embryonic cell cycles and in early mitosis in somatic cells. When there are unattached kinetochores in the cell, Cdc20 is inactivated by the spindle checkpoint to prevent anaphase. APC/C bound to Cdc20 (APC/C^{Cdc20}) seems primarily to recognize substrates with "Destruction box" motifs (see Box 2). In contrast, Cdh1 (*fizzy-related* in *Drosophila*) does not seem to be present in most embryonic cell cycles and is most important for ubiquitination in anaphase and on through the following G1 phase. APC/C bound to Cdh1 (APC/C^{Cdh1}) can recognize substrates with either a D-box or a KEN box and thus has a wider range of substrates than APC/C^{Cdc20}. Ama1 only acts in meiosis and, remarkably, one of the APC/C subunits in mitotic cells inhibits Ama1 to prevent it acting prematurely in meiosis and in mitotic cells (Oelschlaegel et al. 2005; Penkner et al. 2005).

Cdh1 is able to bind and activate the APC/C in interphase and in quiescent or differentiated cells (reviewed in Peters 2002). Therefore, to keep Cdh1 from prematurely activating the APC/C in G2 phase, it is kept inactive by phosphorylation by G2 cyclin-Cdk activity (Lukas et al. 1999; Zachariae et al. 1998). In animal cells, Cdh1 is also sequestered and inactivated by Rca1 in Drosophila or its vertebrate homologue Emi1 (Dong et al. 1997; Grosskortenhaus and Sprenger 2002; Reimann et al. 2001). In the absence of Emi1, cells are unable to inactivate APC/CCdh1 and, therefore, cannot accumulate the mitotic cyclins or the geminin protein that regulate DNA replication. As a result, cells without Emi1 endoreplicate and cannot divide (DiFiore and Pines 2007).

Activating the APC/C and Recognizing its Prometaphase Substrates

When cyclin B-Cdk1 is fully activated, cells are committed to mitosis. At the same time, cells become committed to exit from mitosis through cyclin B-Cdk1 phosphorylating the APC/C, enabling it to bind to Cdc20 and, in turn, recognize the mitotic cyclins, including cyclin B1, as a substrate. The mitotic cyclins were the first APC/C substrates to be characterized, and they get their name from their dramatic instability once cells enter mitosis. On immunoblots or by following radiolabelled proteins, cyclin A always disappears before cyclin B, and in highly synchronous invertebrate eggs the disappearance of cyclin B correlates with anaphase. However, to understand how the degradation of a specific protein is controlled, the key event is when its degradation starts, not when the protein has disappeared. To this end, live cell imaging of fusion proteins made between a protein of interest and green fluorescent protein (GFP) has proved very useful. The fusion proteins act as markers for the UPS-dependent destruction of the endogenous protein because the GFP tag is unfolded and degraded by the proteasome along with the protein to which it is attached. Thus, assaying the amount of fluorescence gives a measure of the amount of protein in the cell.

Using this live cell assay, the earliest APC/C substrate known to date is cyclin A (den Elzen and Pines 2001; Geley et al. 2001). Cyclin A begins to be degraded at, or just after, nuclear envelope breakdown. At present it is unclear why cyclin A has to be degraded so early in mitosis, although expressing an ectopic non-degradable version of cyclin A does prevent *Drosophila* embryo and mammalian tissue culture cells from initiating anaphase (den Elzen and Pines 2001; Geley et al. 2001; Sigrist et al. 1995).

While APC/C^{Cdc20} is busily ubiquitinating cyclin A, the chromosomes are attaching to the spindle, and in somatic cells the spindle assembly checkpoint machinery is activated to prevent the cells from prematurely entering anaphase. This process sets up an as-yet unresolved puzzle because most models of the checkpoint propose that it works by inactivating or sequestering Cdc20 to prevent it from interacting with the APC/C (reviewed in Musacchio and Hardwick 2002). As yet we have no clear idea how the APC/C can still recognize proteins in the presence of the checkpoint. It may be that substrates such as cyclin A are recognized very efficiently by the APC/C, so that even a small amount of Cdc20 is sufficient to promote their degradation, or that they bind to Cdc20 in a manner that prevents the spindle-checkpoint proteins from inactivating Cdc20.

Spindle Checkpoint-dependent APC/C Substrates

The key event in mitosis is the removal of the anaphase and cytokinesis inhibitors only after all of the sister chromatids have correctly attached to the spindle. These inhibitors are securin and cyclin B, respectively, both substrates of APC/C^{Cdc20}. Live cell imaging has revealed that they begin to be degraded at the same time in human cells: when the last unattached kinetochore is captured by a spindle microtubule, i.e., when the spindle checkpoint is inactivated (Clute and Pines 1999; Hagting et al. 2002). Eliminating the spindle checkpoint in somatic cells advances cyclin B and securin destruction to begin at the same time as cyclin A (Hagting et al. 2002). This means that the spindle checkpoint is an integral part of every mitosis in somatic cells; its elimination leads to aneuploidy and is inviable in animal cells.

Although the spindle checkpoint sets the timing for cyclin B and securin destruction in somatic cells, there is a lag between when the APC/C is activated and when cyclin B begins to be degraded even in systems where there is no checkpoint, such as cleaving invertebrate embryos and frog egg extracts. This lag is obviously crucial to keeping the cell in mitosis long enough for the spindle properly to segregate sister chromatids, and at present we don't really know how this works. One mechanism that has been proposed is that Mad2 and BubR1 proteins act independently of their role in the mitotic checkpoint (Meraldi et al. 2004). This proposal is based on siRNA studies in which reducing Mad2 levels in mammalian cultured cells accelerated the average time from nuclear envelope breakdown to anaphase, regardless of whether there were unattached kinetochores. Reducing BubR1 and Mad2 together further accelerated progress to anaphase, indicating that Mad2 and BubR1 might together constitute the timer (Meraldi et al. 2004). However, this model has yet to be tested in an embryonic system.

Separating Sister Chromatids

In the PtK1 rat kangaroo cell line, which has only 11 chromosomes, the capture of the last kinetochores can be accurately assayed and is consistently ~23 min before anaphase (Rieder et al. 1994). This timing raises the questions of what sets the time from chromosome attachment to anaphase – perhaps the time taken to degrade securin and cyclin B – and how do sister chromatids separate synchronously at anaphase?

Sister chromatids are held together by cohesin complexes that assemble during DNA replication (reviewed in Nasmyth 2001). Cohesin complexes are composed of a heterodimer of two structural maintenance of chromosomes (SMC) proteins and two sister chromatid cohesion (Scc) proteins and have been proposed to form a ring that holds sister chromatids together. In vertebrate cells, most of the cohesin complexes on the chromosome arms are removed in prophase by phosphorylation by the Plk and Aurora protein kinases (Losada et al. 2002; Waizenegger et al. 2000), but the complexes at the centromeres are protected by the Shugoshin protein (Kitajima et al. 2005; McGuinness et al. 2005; Salic et al. 2004; Tang et al. 2004; which was originally identified as the protein that protects centromeric complexes from cleavage in Meiosis I in yeast, as reviewed in Watanabe 2005). All the yeast cohesin complexes, and centromeric animal cohesin complexes, are subsequently released by the separase protease that cleaves the Scc1 subunit of the complex. Securin both activates and inhibits separase, and a non-degradable version of securin prevents sister chromatid separation in the yeasts and animal cells (Nasmyth 2001). Therefore, in principle, anaphase could simply be triggered by separase, as it is released from inhibition by securin. However, it is difficult to reconcile the almost simultaneous separation of all sister chromatids with the gradual increase in free separase over, for example, the 20-min period in which securin is released in PtK1 cells. One explanation might be that there is a second signal that triggers the final separation of the chromosomes. In budding yeast, the Polo-like kinase orthologue, Cdc5, must phosphorylate Scc1 to make it a substrate for separase (Alexandru et al. 2001), and this might be coordinated on all chromosomes at the same time. However, this mechanism is not apparently conserved in animal cells where separase does not require a phosphorylated form of cohesin as its substrate; instead Plk1 helps cohesin subunits to disassemble in prophase (Waizenegger et al. 2000). An obvious candidate to impose synchrony in animal cells is the Shugoshin protein that protects centromeric cohesion, but as yet it is unclear how Shugoshin is inactivated and when. Shugoshin does appear to be a substrate of the APC/C, but the exact time at which it is degraded is not known (Salic et al. 2004).

Another mechanism that could provide a second signal has been observed in *Xenopus* egg extracts. Here, the separase protein is phosphorylated and inhibited by mitotic cyclin-Cdk activity (Stemmann et al. 2001), Therefore, separase can only cleave the centromeric cohesin complexes when mitotic Cyclin-Cdk activity falls below a certain threshold, which may also explain why securin is not essential in mammalian cells (Mei et al. 2001). However, this threshold appears to be set fairly high because it requires 1.5- to 2-fold more cyclin-Cdk activity than is present in normal mitotic cells to inhibit separase in *Xenopus* extracts (Stemmann et al. 2001) or in living mammalian cells (Hagting et al. 2002). Early in mitosis, cyclin A-Cdk activity could contribute to inhibiting separase, and DNA damage does delay anaphase in *Drosophila* embryos through stabilizing cyclin A (Su and Jaklevic 2001). However, since cyclin A is mostly degraded by metaphase, it cannot explain how sister chromatid cohesion can persist in cells lacking securin when they are arrested in metaphase.

Cut Phenotypes

Although cyclin B and securin are degraded at the same time in animal cells and are under the same control by the spindle checkpoint, they are not dependent on one another. A non-degradable version of securin will prevent sister chromatid separation, but cyclin B proteolysis continues on-schedule, such that the cell attempts cytokinesis in the presence of unseparated chromosomes. This generates a "cut" (cell untimely torn) phenotype in fission yeast cells where the septum divides the nucleus (Yanagida 1998). In animal cells, the outcome is variable depending on the position of the chromosomes in the cell; sometimes all of the chromosomes are partitioned into only one daughter cell, in other cases the cleavage furrow attempts to divide the chromosome mass and, in most cases, eventually regresses to generate one tetraploid cell. Large numbers of fission yeast "cut" mutants have been isolated, and it is interesting to note that some of these are mutations in different APC/C subunits, which may be evidence that particular interaction domains on specific APC/C subunits have a role in recognizing different substrates. Alternatively, some substrates may be recognized at higher affinity than others.

Spatial Control of Proteolysis

Remarkably, the spindle assembly checkpoint can rapidly inactivate cyclin B1 and securin proteolysis even after it has begun. Adding taxol or nocodazole to metaphase cells arrests them because the drugs re-impose the spindle checkpoint and turn off cyclin B1 and securin destruction (Clute and Pines 1999; Hagting et al. 2002). When this experiment is performed with taxol in mammalian cells, there is a striking re-localization of cyclin B1 to the spindle poles and chromosomes, indicating that cyclin B1 may need to flux onto the spindle to be degraded (Clute and Pines 1999). In agreement with this finding, although the bulk of the population of cyclin B1 is not degraded in the cell division cycles of Drosophila embryos, a sub-population around the spindle is destroyed, which is required for cells to enter anaphase (Huang and Raff 1999). In these embryos, a wave of cyclin B1 proteolysis appears to begin at the centrosomes and spread to the middle of the spindle. Furthermore, in mutant embryos where the centrosomes detach from the spindle, cyclin B1 is degraded on the detached centrosome but not on the rest of the spindle (Wakefield et al. 2000). These experiments indicate that cyclin B1 ubiquitination may be spatially regulated in cells, and the phenotype of embryos lacking the Drosophila UBC10 family member, vihar, and the localization of the vihar protein to the spindle and spindle poles indicate that some of the spatial control on cyclin B destruction may be orchestrated by vihar. Immunofluorescence studies in Drosophila and mammalian cells have revealed that the APC/C is localized to the spindle, in particular to the spindle poles (Kraft et al. 2003; Acquaviva et al. 2004), and, in prophase and pro-metaphase, to unattached kinetochores (Acquaviva et al. 2004). Possibly the ubiquitination of APC/C^{Cdc20} substrates is spatially regulated to facilitate the close coupling between the spindle checkpoint and the APC/C (reviewed in Pines and Lindon 2005).

Leaving Mitosis

In somatic cells, the decline in cyclin B-Cdk activity allows the APC/C to bind Cdh1. In budding yeast, the Cdc14 phosphatase is responsible for dephosphorylating Cdh1, but it is unclear whether this is true in animal cells. The result of binding Cdh1 is that the APC/C now recognizes a wider set of substrates: those with D-boxes and those with KEN boxes. One of these substrates is Cdc20 itself (Pfleger and Kirschner 2000), meaning that there is a complete switch from APC/C^{Cdc20} to APC/C^{Cdh1}, and one consequence of this is that the spindle assembly checkpoint machinery can no longer turn off the APC/C. The proteins targeted by APC/C^{Cdh1} include regulatory proteins, such as the mitotic kinases, and geminin, an inhibitor of DNA replication, whose ubiquitination – but not necessarily destruction – allows cells to re-license origins of replication as they re-enter interphase (Li and Blow 2004). They also include proteins that are functional components of mitosis- or cytokinesis-specific structures, such as the mitotic spindle, cytokinetic furrow (Zhao and Fang 2005), and kinetochores, which must be disassembled to return the cell to its interphase state.

The mitotic regulators targeted by APC/C^{Cdh1} include Cdc5/Plk1 and the Aurora A kinase (Castro et al. 2002; Lindon and Pines 2004; Littlepage and Ruderman 2002; Shirayama et al. 1998). Live cell imaging reveals that these proteins are degraded at different times in anaphase, indicating that there are further controls on the timing of when Cdh1 can recognize its substrates (Lindon and Pines 2004). For Aurora A, this may be through modification of a second motif, the A box or D-box activating domain, that is required for Aurora A to be ubiquitinated and whose phosphorylation inhibits destruction in vitro (Castro et al. 2002; Littlepage and Ruderman 2002). Both these protein kinases can also be inactivated by alternative pathways, such as dephosphorylation or, for Aurora A, dissociation from its activating partner TPX2 mediated by the p97 AAA-ATPase, which is required for spindle disassembly in *Xenopus* extracts (Cao et al. 2003). Thus, proteolysis is not essential to inactivate them but it does appear to promote efficient mitotic exit. For example, a non-degradable version of Plk1 perturbs cytokinesis and interferes with coordination between the position of the cleavage furrow and the mitotic spindle (Lindon and Pines 2004). Indeed, it appears that none of the APC/C^{Cdh1} substrates must be degraded for cells to exit from mitosis (Jacobs et al. 2002); the most profound effects in cells lacking Cdh1 are on the regulation of events and decisions in G1 phase, in maintaining quiescence (Wirth et al. 2004) and in post-mitotic cells (Peters 2002).

Concluding Remarks

Ubiquitin-mediated proteolysis is a rapid and decisive mechanism to control progress through mitosis, to aid in cytokinesis, and to return cells to their interphase state. In budding yeast, cytokinesis requires another ubiquitin ligase, SCF^{Grr1}, which is recruited to the region of the mother-bud neck where it binds and degrades the Hof1/Cyk2 protein to allow the efficient contraction of the acto-myosin ring (Blondel et al. 2005). Genetic screens in the yeasts and *C. elegans* have indicated that other ubiquitin ligases may also be involved in regulating mitosis (Hermand et al. 2003; Michel et al. 2003), although in each case it is important to determine whether these are direct effects or the consequences of entering mitosis with damaged or unreplicated DNA. In animal cells, conditional knockouts of the core APC subunits, APC2 and APC11, have revealed an important role in maintaining cells in their quiescent state, and recent evidence from invertebrate systems has indicated a role in synaptic plasticity. In mitosis, the APC/C has the crucial role in selecting the right substrate at the right time, in part through associating with different WD40 proteins at different times, but elucidating exactly how it selects its substrates and how it responds to the spindle checkpoint will be essential to a proper understanding of how mitosis is regulated.

References

Acquaviva C, Herzog F, Kraft C, Pines J (2004) The anaphase promoting complex/cyclosome is recruited to centromeres by the spindle assembly checkpoint. Nature Cell Biol 6:892–898

Alexandru G, Uhlmann F, Mechtler K, Poupart MA, Nasmyth K (2001) Phosphorylation of the cohesin subunit Scc1 by Polo/Cdc5 kinase regulates sister chromatid separation in yeast. Cell 105:459–472

Blondel M, Bach S, Bamps S, Dobbelaere J, Wiget P, Longaretti C, Barral Y, Meijer L, Peter M (2005) Degradation of Hof1 by SCF(Grr1) is important for actomyosin contraction during cytokinesis in yeast. Embo J 24:1440–1452

Cao K, Nakajima R, Meyer HH, Zheng Y (2003) The AAA-ATPase Cdc48/p97 regulates spindle disassembly at the end of mitosis. Cell 115:355–367

Carroll CW, Morgan DO (2002) The Doc1 subunit is a processivity factor for the anaphase-promoting complex. Nature Cell Biol 4:880–887

Carroll CW, Enquist-Newman M, Morgan DO (2005) The APC subunit Doc1 promotes recognition of the substrate destruction box. Curr Biol 15:11–18

Castro A, Vigneron S, Bernis C, Labbe JC, Prigent C, Lorca T (2002) The D-Box-activating domain (DAD) is a new proteolysis signal that stimulates the silent D-Box sequence of Aurora-A. EMBO Rept 3:1209–1214

Clute P, Pines J (1999) Temporal and spatial control of cyclin B1 destruction in metaphase. Nature Cell Biol 1:82–87

den Elzen N, Pines J (2001) Cyclin A is destroyed in prometaphase and can delay chromosome alignment and anaphase. J Cell Biol 153:121–136

DiFiore B, Pines J (2007) J Cell Biol 177:425–437

Dong X, Zavitz KH, Thomas BJ, Lin M, Campbell S, Zipursky SJ (1997) Control of G1 in the developing Drosophila eye: rca1 regulates Cyclin A. Genes Dev 11:94–105

Elia AE, Cantley LC, Yaffe MB (2003) Proteomic screen finds pSer/pThr-binding domain localizing Plk1 to mitotic substrates. Science. 299:1228–1231

Geley S, Kramer E, Gieffers C, Gannon J, Peters JM, Hunt T (2001) APC/C-dependent proteolysis of human cyclin A starts at the beginning of mitosis and is not subject to the spindle assembly checkpoint. J Cell Biol 153:137–148

Grosskortenhaus R, Sprenger FR (2002) Rca1 inhibits APC-Cdh1(Fzr) and is required to prevent cyclin degradation in G2. Dev Cell 2:29–40

Hagting A, Den Elzen N, Vodermaier HC, Waizenegger JC, Peters JM, Pines J (2002) Human securin proteolysis is controlled by the spindle checkpoint and reveals when the APC/C switches from activation by Cdc20 to Cdh1. J Cell Biol 157:1125–1137

Hermand D, Bamps S, Tafforeau L, Vandenhaute J, Makela TP (2003) Skp1 and the F-box protein Pof6 are essential for cell separation in fission yeast. J Biol Chem 278:9671–9677

Hershko A, Ciechanover A (1998) The ubiquitin system. Annu Rev Biochem 67:425–479

Huang J, Raff JW (1999) The disappearance of cyclin B at the end of mitosis is regulated spatially in Drosophila cells. Embo J 18:2184–2195

Jacobs H, Richter D, Venkatesh D, Lehner C (2002) Completion of mitosis requires neither fzr/rap nor fzr2, a male germline-specific Drosophila Cdh1 homolog. Curr Biol 12:1435–1441

Kitajima TS, Hauf S, Ohsugi M, Yamamoto T, Watanabe Y (2005) Human Bub1 defines the persistent cohesion site along the mitotic chromosome by affecting Shugoshin localization. Curr Biol 15:353–359

Kraft C, Herzog F, Gieffers C, Mechtler K, Hagting A, Pines J, Peters JM (2003) Mitotic regulation of the human anaphase-promoting complex by phosphorylation. Embo J 22:6598–6609

Li A, Blow JJ (2004) Non-proteolytic inactivation of geminin requires CDK-dependent ubiquitination. Nature Cell Biol 6:260–267

Lindon C, Pines JC (2004) Ordered proteolysis in anaphase inactivates Plk1 to contribute to proper mitotic exit in human cells. J Cell Biol 164:233–241

Littlepage LE, Ruderman JV (2002) Identification of a new APC/C recognition domain, the A box, which is required for the Cdh1-dependent destruction of the kinase Aurora-A during mitotic exit. Genes Dev 16:2274–2285

Losada A, Hirano M, Hirano T (2002) Cohesin release is required for sister chromatid resolution, but not for condensin-mediated compaction, at the onset of mitosis. Genes Dev 16:3004–3016

Lukas C, Sorensen CS, Kramer E, Santoni-Rugiu E, Lindeneg C, Peters JM, Bartek J, Lukas J (1999) Accumulation of cyclin B1 requires E2F and cyclin-A-dependent rearrangement of the anaphase-promoting complex. Nature 401:815–818

McGuinness BE, Hirota T, Kudo NR, Peters JM, Nasmyth K (2005) Shugoshin prevents dissociation of cohesin from centromeres during mitosis in vertebrate cells. PLoS Biol 3:e86

Mei J, Huang X, Zhang P (2001) Securin is not required for cellular viability, but is required for normal growth of mouse embryonic fibroblasts. Curr. Biol 11:1197–1201

Meraldi P, Draviam VM, Sorger PK (2004) Timing and checkpoints in the regulation of mitotic progression. Dev Cell 7:45–60

Michel JJ, McCarville JF, Xiong Y (2003) A role for Saccharomyces cerevisiae Cul8 ubiquitin ligase in proper anaphase progression. J Biol Chem 278:22828–22837

Mimura S, Seki T, Tanaka S, Diffley JF (2004) Phosphorylation-dependent binding of mitotic cyclins to Cdc6 contributes to DNA replication control. Nature 431:1118–1123

Musacchio A, Hardwick KG (2002) The spindle checkpoint: structural insights into dynamic signalling. Nature Rev Mol Cell Biol 3:731–741

Nasmyth K (2001) Disseminating the genome: joining, resolving, and separating sister chromatids during mitosis and meiosis. Annu Rev Genet 35:673–745

Nigg EA (2001) Mitotic kinases as regulators of cell division and its checkpoints. Nature Rev Mol Cell Biol 2:21–32

Oelschlaegel T, Schwickart M, Matos J, Bogdanova J, Camasses A, Havlis J, Shevchenko A, Zachariae W (2005) The yeast APC/C subunit Mnd2 prevents premature sister chromatid separation triggered by the meiosis-specific APC/C-Ama1. Cell 120:773–788

Passmore LA (2004) The anaphase-promoting complex (APC): the sum of its parts? Biochem Soc Trans 32:724–727

Passmore LA, McCormack EA, Au SW, Paul A, Willison KR, Harper JW, Barford D (2003) Doc1 mediates the activity of the anaphase-promoting complex by contributing to substrate recognition. Embo J 22:786–796

Penkner AM, Prinz S., Ferscha S, Klein F (2005) Mnd2, an essential antagonist of the anaphase-promoting complex during meiotic prophase. Cell 120:789–801

Peters JM (2002) The anaphase-promoting complex: proteolysis in mitosis and beyond. Mol Cell 9:931–943

Pfleger CM, Kirschner MW (2000) The KEN box: an APC recognition signal distinct from the D box targeted by Cdh1. Genes Dev 14:655–665

Pines J, Rieder CJ (2001) Re-staging mitosis: a contemporary view of mitotic progression. Nature Cell Biol 3:E3–E6

Pines J, Lindon C (2005) Proteolysis: anytime, any place, anywhere? Nature Cell Biol 7:731–735

Reimann JD, Freed E, Hsu JY, Kramer E, Peters JM, Jackson PK (2001) Emi1 is a mitotic regulator that interacts with Cdc20 and inhibits the anaphase promoting complex. Cell 105:645–655

Rieder CL, Schultz A, Cole R, Sluder G (1994) Anaphase onset in vertebrate somatic cells is controlled by a checkpoint that monitors sister kinetochore attachment to the spindle. J Cell Biol 127:1301–1310

Rudner AD, Murray AW (2000) Phosphorylation by Cdc28 activates the Cdc20-dependent activity of the anaphase-promoting complex. J Cell Biol 149:1377–1390

Salic A, Waters JC, Mitchison TJ (2004) Vertebrate shugoshin links sister centromere cohesion and kinetochore microtubule stability in mitosis. Cell 118:567–578

Shirayama M, Zachariae W, Ciosk R, Nasmyth K (1998) The Polo-like kinase Cdc5p and the WD-repeat protein Cdc20p/fizzy are regulators and substrates of the anaphase promoting complex in Saccharomyces cerevisiae. EMBO J 17:1336–1349

Sigrist S, Jacobs H, Stratmann R, Lehner CF (1995) Exit from mitosis is regulated by Drosophila fizzy and the sequential destruction of cyclins A, B and B3. EMBO J 14:4827–4838

Stemmann O, Zou H, Gerber SA, Gygi SP, Kirschner MW (2001) Dual inhibition of sister chromatid separation at metaphase. Cell 107:715–726

Su TT, Jaklevic B (2001) DNA damage leads to a Cyclin A-dependent delay in metaphase-anaphase transition in the Drosophila gastrula. Curr Biol 11:8–17

Tang Z, Sun Y, Harley SE, Zou H, Yu H (2004) Human Bub1 protects centromeric sister-chromatid cohesion through Shugoshin during mitosis. Proc Natl Acad Sci USA. 101:18012–18017

Vodermaier HC (2001) Cell cycle: waiters serving the destruction machinery. Curr Biol 11:R834–R837

Vodermaier HC, Gieffers C, Maurer-Stroh S, Eisenhaber F, Peters JM (2003) TPR subunits of the anaphase-promoting complex mediate binding to the activator protein CDH1. Curr Biol 13:1459–1468

Waizenegger IC, Hauf S, Meinke A, Peters JM (2000) Two distinct pathways remove mammalian cohesin from chromosome arms in prophase and from centromeres in anaphase. Cell 103:399–410

Wakefield JG, Huang J, Raff JW (2000) Centrosomes have a role in regulating the destruction of cyclin B in early drosophila embryos. Curr Biol 10:1367–1370

Watanabe Y (2005) Sister chromatid cohesion along arms and at centromeres. Trends Genet 21:405–412

Wirth KG, Ricci R, Gimenez-Abian JF, Taghybeeglu S, Kudo NR, Jochum S, Vasseur-Cognet M, Nasmyth K (2004) Loss of the anaphase-promoting complex in quiescent cells causes unscheduled hepatocyte proliferation. Genes Dev 18:88–98

Yanagida M (1998) Fission yeast cut mutations revisited: control of anaphase. Trends Cell Biol. 8:144–149

Zachariae W, Schwab M, Nasmyth K, Seufert W (1998) Control of cyclin ubiquitination by CDK-regulated binding of Hct1 to the anaphase promoting complex. Science 282:1721–1724

Zhao WM, Fang G (2005) Anillin is a substrate of APC/C that controls spatial contractility of myosin during late cytokinesis. J Biol Chem 280:33516–33524

Nuclear Receptors and Cyclins in Hormone Signaling

Michael J. Powell[1], *Vladimir M. Popov*[1], *Chenguang Wang*[1], and *Richard G. Pestell*[1]

Abstract

Hormones were previously known to induce kinase signaling cascades from the cell surface to the nucleus. Nuclear receptors (NR) coordinate diverse biological phenotypes altering cellular metabolism, differentiation and cellular proliferation. The genes that regulate cell-cycle progression participate as targets of NR signaling and coordinate many of these responses. These cyclins and CDKs in turn feed back to modify NR activity. Proteins governing nuclear receptor (estrogen receptor α (ERα), androgen receptor (AR), TR, GR) function can be modified by acetylation. Such proteins include the co-activator (SRC1), co-integrator (p300, p/CAF), HSP-90, histones and HDACs. Phosphorylation, acetylation and ubiquitination occur within transcription factors as a form of intra-transcription factor signaling (signaling cascades within transcription factors; SCITS). The modification of nuclear receptors by acetylation determines gene expression specificity and is a key determining factor of cellular growth in hormone-responsive cells.

Cell Cycle Control

The orderly progression of cells through the cell cycle is temporally coordinated, monitored and executed by cyclins and CDK enzyme complexes. The holoenzymes that participate in coordinating cell cycle progression are highly conserved between species and consist of regulatory and catalytic subunits capable of phosphorylating a subset of key target substrates. More than 13 CDKs and 25 proteins with homology in the cyclin box domain have been identified in the human genome (Liu et al. 2004). Irrevocable commitment of the cell to synthesize DNA is monitored during the mid and late G1 phase of the cell cycle. The favorability of the local micro- and macro-environment is sensed during this transition step. Execution of this transition is primarily coordinated by the D-type cyclins and sequentially by cyclin E in the late G1 phase.

The cyclin regulatory subunit heterodimerizes with either its catalytic partner or its respective CDK to form holoenzymes. This forms part of a complex that phosphorylates and inactivates (in the case of the D-type cyclins) the pRb tumor suppressor. The abundance of cyclin D1 is rate-limiting in cellular proliferation in a variety of different cell types, including breast epithelial cells and other hormone-responsive tissues. The

[1] Kimmel Cancer Center, Department of Cancer Biology, Thomas Jefferson University, Philadelphia, PA, 233 S. 10th Street, Room 1050 Philadelphia PA 19107, USA

e-mail:richard.pestell@jefferson.edu

Melmed et al.
Hormonal Control of Cell Cycle
© Springer-Verlag Berlin Heidelberg 2008

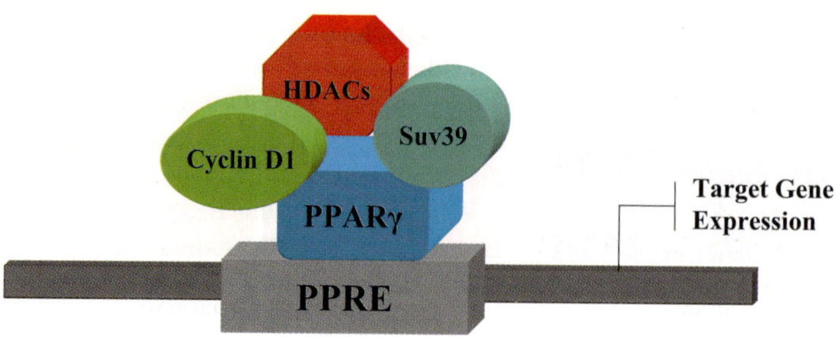

Fig. 2. Cyclin D1 recruits chromatin remodeling complexes

ing transgenic and knockout mice. The studies revealed that cyclin D1 could recruit a subset of histone-modifying proteins to the DNA-binding site of the PPARγ nuclear receptor. The recruitment of cyclin D1 in the context of local chromatin revealed the co-recruitment of histone deacetylases, in particular HDAC1 and HDAC3. Cyclin D1 recruitment of HDACs correlated with the deacetylation of local histones surrounding the PPARγ response element. The deacetylation of histone H3 at lysine 9 was associated with the recruitment of cyclin D1. Collectively, these studies demonstrate that cyclin D1 plays a modifying role in deacetylating histones. In addition, the cyclin D1 recruitment to an endogenous PPARγ element was associated with the recruitment of HP1 and the histone methylase SUV39 (Fu et al. 2005a,b; Fig. 2). These studies were significant since they demonstrated that, in addition to regulating phosphorylation of Rb, cyclin D1 regulated the recruitment of HDACs. Histone deacetylation is thought to play an important role in the tumorigenic phenotype regulated by genetic breakpoint rearrangements in a variety of hematological and solid tumors. These studies raise the question of whether the ability of cyclin D1 to recruit histone deacetylases may contribute to the cyclin D1-mediated collaborative oncogenic phenotype.

Nuclear Receptor Acetylation

NRs share a subset of conserved domains and bind distinct ligands, including steroids, thyroid hormone, retinoids, vitamins, and as yet undefined ligands. Nuclear receptors coordinate physiological homeostasis, reproduction and cellular metabolism. Nuclear receptor conserved domains include the activation function domains, hinge region, DNA-binding domain and ligand-binding domain. Through these domains, nuclear receptors form coordinated inter- and intra-molecular interactions with associated proteins and enzymes to render the previously mentioned functions. Historically, research has focused on the signaling cascades that emanate from the cell surface and through intracellular signaling, regulate nuclear receptor function. In this model, the NR is seen as a destination molecule conducting the outcome of the cytoplasmic signaling pathway. More recently, the complexity of the protein-protein interactions of nuclear receptors, both within the cytoplasm and the nucleus, have identified important modifiers of these signaling pathways at the level of nuclear receptor bound proteins. Thus, nuclear receptor binding proteins, such as SRC-1, SRC-2, SRC-3, p300/CBP-P/CAF, and nuclear repressors, including NCoR, SMRT, Sin3, HDAC, BRC-1, and NURD, can modify the outcome of cellular signaling pathways. More recently, it has become clear that signaling pathways exist within the nuclear receptor themselves. The nuclear receptors undergo posttranslational modification by phosphorylation, acetylation and methylation. These modifications are thought to create a mechanism for signaling cascades within the transcription factor, which in turn may define the distinct genetic signaling output of the nuclear receptor.

The notion that histone acetyltransferases may regulate hormone signaling began with the observation that epigenomic modifications create reversible, inheritable chromatin alterations. DNA methyltransferases identified in 1983 provided evidence of methyltransferases that alter DNA structure (Feinberg and Vogelstein 1983). The identification of histone acetyltransferases (Kleff et al. 1995) provided a tractable genetic mechanism through which histone modification may regulate cellular differentiation. Subsequent identification of enzymes involved in histone acetylation, DNA methylation and histone phosphorylation, ubiquitination and methylation has given rise to the notion that each of these posttranslational alterations are tightly coordinated to regulate signal transduction and specificity. More recently, the characteristic changes in these posttranslational modifications of histones have lead to the understanding that these modifications encode specific signaling transduction pathways that are altered in response to hormone signaling. Strahl and Allis proposed that posttranslational histone modifications encode "a histone code" (Strahl and Allis 2000).

The observations that nuclear receptors are directly acetylated (Fu et al. 2000) and that the site of acetylation is conserved between nuclear receptors (Wang et al. 2001) have led to a new area of research. Subsequent studies identified functional acetylation sites within a number of nuclear receptors. The androgen receptor is acetylated within the hinge region by several histone acetyltransferases, including P/CAF, p300 and TIP60. Of particular importance, however, was the observation that a nuclear receptor acetylation site was a key regulator of cellular growth, in particular contact-independent growth. In the past, phosphorylation of target substrates was known to regulate cellular growth. With the finding that single point substitutions within the acetylation site of the androgen receptor could function as a molecular switch to convert the androgen receptor to a promoter of contact-independent growth, a new mechanism was identified for targeted therapeutics. Prostate cancer cells, transduced with a gain-of-function androgen receptor acetylation site mutant, promoted the growth of human prostate cancer cells (Bouras et al. 2005). As originally predicted, numbers of nuclear receptors are now known to be directly acetylated, including the ERα, AR, TRβ and GR (Table 1).

As our understating has grown, it has become clear that, just as phosphorylation occurs at multiple distinct components of the signaling cascades to regulate hormone signaling, so too acetylation occurs at multiple distinct components throughout hormone signaling cascades (Fig. 3). In this regard, nuclear chaperone proteins, such as HSP90, kinases (MEKK, IKK), NR co-activators (ACTR, p300) and repressors (HDAC), which are proven to regulate hormone signaling, are also regulated by acetylation (Fig. 3; Table 1). A growing body of data now suggests that the acetylation site of nuclear receptors modulates other posttranslational modifications including phosphorylation, ubiquitination and sumoylation. Thus, ERα and AR acetylation modulates phosphorylation and signaling specificity. The AR acetylation site regulates Akt- but not cAMP-mediated activity. The concept then is that nuclear receptors contain within themselves, intramolecular signaling cascades. Similar observations have been made

Table 1. FAT substrates

Substrates for FAT	FAT	Possible effects on transcription
General transcriptional factors		
TFIIF	p300/CBP, P/CAF, TIP60	Up
TFIIEB	p300/CBP, P/CAF, TAFII250	Unknown
TAF(I)68	P/CAF	Up
UBF	CBP	Up
CIITA	P/CAF	Up
Nuclear Receptors		
AR	p300/CBP, P/CAF	Up (Fu et al. 2000, 2003)
ERα	p300	Up (Fu et al. 2003; Wang et al. 2001)
GR	HDAC2	(Ito et al. 2006)
TRβ		(Lin et al. 2005)
Transcriptional effectors		
β-catenin	CBP	(Wolf et al. 2002)
CEBPβ	p300	Activation (Cesena et al. 2007)
C-Myb	p300/CBP, GCN5	Up
E1A	p300/CBP	Up
E2F, 4	TRRAP, P/CAF	Up
E2F1, 2	p300/CBP, P/CAF	Up
EKLF	p300/CBP	Up
GATA-1,-3	p300/CBP, PCAF	Up
HIF-1α	ARD1	Degradation (Jeong et al. 2002)
HIV-1 tat	p300/CBP, P/CAF	Up
HSP90		(Scroggins et al. 2007)
IRF-1, 2	p300/CBP, P/CAF	Up
MyoD	p300/CBP, P/CAF	Up
NF-E2	p300/CBP	Up
P53	p300/CBP, P/CAF	Up
P65	p300	Up (Ishinaga et al. 2007)
pRb	p300/CBP	Up
RelA	p300/CBP	Nuclear import
SF-1	GCN5	Up
SMAD3	p300/CBP	Up (Inoue et al. 2007; Simonsson et al. 2006)
Spl	p300/CBP	Up
STAT6	p300/CBP	Up
TAL1/SCL	p300/CBP, P/CAF	Up
TCF	p300/CBP	Down
TR-RXR	p300/CBP	
YY1	p300/CBP, P/CAF	Down
Nuclear receptor coactivators		
p300/CBP	p300/CBP	Activation (Bouras et al. 2005)
P/CAF	P/CAF	Unknown
ACTR	p300/CBP	Down
SRC-1	p300/CBP	Unknown
TIF2	p300/CBP	Unknown
Rip140	p300/CBP	Up
PC4	p300	Up
MTA1	p300	Up (Gururaj et al. 2006)
Nonhistone chromatin proteins		
HMG1	p300/CBP	(Pasheva et al. 2004)
HMG2	—	(Pasheva et al. 2004)
HMG14	p300/CBP	Down
HMG17	P/CAF	Unknown
HMG1 (Y)	p300/CBP, P/cAF	Up (P/CAF), Down (p300)
Sin1	GCN5	Unknown
Fen-1	p300	Reduce DNA binding and nuclease activity
Others		
α-Tubulin	p300	Unknown
Importin-α	p300/CBP	Unknown
CDP/cut	p300/CBP, P/CAF	Reduce DNA binding
MEK2, IKKα, β	YopJ	Inhibition (Mittal et al. 2006)

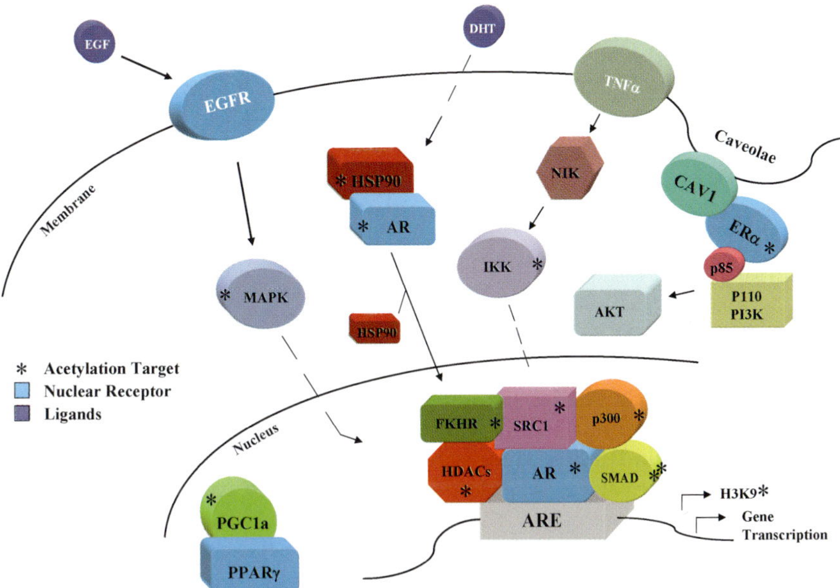

Fig. 3. Hormone cellular signaling cascades and acetylation. The signaling cascade that regulates hormone signaling contains multiple components, including lipoproteins and kinases. Each of the components indicated by an asterisk is now known to be acetylated. These studies suggest that acetylation of NR and the additional components of the hormone signaling pathway play a key role in hormone signaling specificity

with other transcription factors and histones, in which modification by phosphorylation determines subsequent acetylation and activity of the transcription factor. Collectively these findings suggest a role for intramolecular receptor signaling pathways in fine-tuning of gene expression. Such intermolecular signaling within nuclear receptors is referred to as SCITS (signal cascade within transcription factor). Evidence supporting this model includes the finding that phosphorylation and acetylation regulates NR chromatin access at targeted DNA binding sites. Mutation at NR acetylation sites modifies NR access in the context of local chromatin. Distinct acetylation sites of p53 have been shown to contribute to the distinct signal transduction pathways (Knights et al. 2006). Androgen receptor acetylation has also been shown to regulate resistance to the hormone antagonist flutamide (Fu et al. 2003). The AR acetylation site regulates cellular growth through both the inhibition of apoptosis and the induction of cellular proliferation. Additionally, the AR acetylation site mutants induce a subset of cell cycle regulatory genes, including cyclin D1, to promote prostate cellular growth.

Recent studies have shown that the NAD-dependent histone deacetylases (Sirtuins) regulate NR function in an acetylation site-specific manner. The AR is a substrate for SirT1. SirT1 inhibits ligand-dependent activation of the AR, requiring the catalytic function of Sirt1 (Fu et al. 2006). The AR and SirT1 are co-associated in cultured cells. ERα and PPARγ activities are regulated by chemical inhibition of the Sirtuins, raising the possibility that these NRs also serve as substrates for SirT1. The NR co-activator, p300, is deacetylated by SirT1 (Bouras et al. 2005). As the relative abundance of p300 is often rate-limiting in NR and transcription factor activity, and Sirt1 function is dependent upon local NAD/NADH concentration, it is likely that Sirt1-mediated deacetylation of p300 plays a dynamic role in coordinating metabolic gene expression and responses.

Summary

Cyclins encode regulatory subunits of holoenzymes that phosphorylate pRB family members. The abundance of cyclin D1 regulates diverse functions, including the G_1 phase of the cell-cycle, cellular migration, lipid metabolism, mitochondrial function and oncogenic transformation (Albanese et al. 1995; Pestell et al. 1999; Wang et al. 2006; Fig. 1). Cyclin D1 abundance is regulated directly by oncogenic, growth and nutritional signaling pathways. In addition to the pRb family, cyclins bind and regulate the activity of nuclear receptors. The NR superfamily consists of conserved modular transcriptional regulators. The "classical" receptor subclass is comprised of an N-terminal region, Activation Function 1 (AF1), a well-conserved central DNA binding domain with two zinc finger domains and a C-terminal region that includes the hinge and ligand-binding domain (LBD). Distinct types of functional interactions between cyclins and nuclear receptors coordinate metabolism, cellular differentiation and pro-liferation. The expression and abundance of cyclins is regulated by NRs, and NRs in turn regulate the expression and activity of cyclins. Studies of *cyclin D1* knockout mice and of tissue-specific inducible transgenic mice suggest an important role for cyclin D1 in NR function and metabolism in vivo. Cyclin-dependent kinases phosphorylate NRs, whereas the kinase-independent functions of cyclins are capable of regulating the activity of several nuclear receptors (i.e., AR, ERα and PPARγ). Cyclin D1 me-diates the recruitment of histone acetylases (HATs) and histone deacetylases (Suv39, HP1) to modulate histone acetylation in the context of local chromatin at NR binding sites.

NR activity is regulated by phosphorylation, ubiquitination, and acetylation. Acety-lation of NRs regulates cellular growth. NRs (ERα, AR, TR, GR and others), are acety-lated at a motif that is conserved between species and other NRs (Wang et al. 2001). Acetylation of the AR and ERα occurs in cultured cells, whereas point mutations at the acetylation site have been identified in the breast and prostate. The AR and ERα are regulated by TSA-sensitive HDACs and NAD-dependent HDACs (Sirtuins; Fu et al. 2006). The NR acetylation site governs ligand sensitivity, hormone antagonist re-sponses, binding of co-activators and co-repressors (NCoR/HDAC/Smad) and growth properties of the receptors in vivo. The enhanced growth properties of AR acetylation mimics correlate with altered regulation of a select subset of promoters for cell-cycle target genes, including the cyclins.

References

Albanese C, Johnson J, Watanabe G, Eklund N, Vu D, Arnold A, Pestell RG (1995) Transforming p21^{ras} mutants and c-Ets-2 activate the cyclin D1 promoter through distinguishable regions. J Biol Chem 270:23589–23597

Bouras R, Fu M, Sauve AA, Wang F, Quong AA, Perkins ND, Hay RT, Gu W, Pestell RG (2005) SIRT1 deacetylation and repression of P300 involves lysine residues 1020/1024 within the cell-cycle regulatory domain 1. J Biol Chem 280:10264–10276

Cesea RI, Cardinaux JR, Kwok R, Schwartz J (2007)CCAAT/enhancer-binding protein (C/EBP) beta is acetylated at multiple lysines: acetylation of C/EBPbeta at lysine 39 modulates its ability to activate transcription. J Biol Chem 282:956–967

Feinberg AP, Vogelstein B (1983) Hypomethylation distinguishes genes of some human cancers from their normal counterparts. Nature 301:89–92

Fu M, Wang C, Reutens AT, Wang J, Angeletti RH, Siconolfi-Baez L, Ogryzko V, Avantaggiati ML, Pestell RG (2000) p300 and p300/cAMP-response element-binding protein-associated factor acetylate the androgen receptor at sites governing hormone-dependent transactivation. J Biol Chem 275:20853–20860

Fu M, Wang C, Wang J, Sakamaki T, Di Vizio D, Zhang X, Albanese C, Balk S, Chang C, Fan S, Rosen E, Palvimo JJ, Janne OA, Muratoglu S, Avantaggiati ML, Pestell RG (2003) Acetylation of the androgen receptor enhances coactivator binding and promotes prostate cancer cell growth. Mol Cell Biol 23:8563–8575

Fu M, Wang, C, Li Z, Sakamaki T, Pestell RG (2004) Minireview: Cyclin D1: normal and abnormal functions. Endocrinology 145:5439–5447

Fu M, Rao M, Bouras T, Wang C, Wu K, Zhang X, Li Z, Yao T-P, Pestell RG (2005a) Cyclin D1 inhibits PPARgamma-mediated adipogenesis through HDAC recruitment. J Biol Chem 280:16934–16941

Fu M, Wang C, Rao M, Wu X, Bouras T, Zhang X, Li Z, Jiao X, Yang J, Li A, Perkins ND, Thimmapaya B, Kung AL, Munoz A, Giordano A, Lisanti MO, Pestell RG (2005b) Cyclin D1 represses p300 transactivation through a CDK-independent mechanism. J Biol Chem 28:29728–29742

Fu M, Liu M, Sauve AA, Jiao X, Zhang X, Wu X, Powelll MJ, Yang T, Gu W, Avantaggiati ML, Pattabiraman N, Pestell TG, Wang F, Quong AA, Wang C, Pestell RG (2006) Hormonal control of androgen receptor function through SIRT1. Mol Cell Biol 26:8122–8135

Gururai AD, Singh RR, Rayala SK, Homl C, den Hollander P, Zhang H, Balasenthil S, Talukder AH, Landberg G, Kumar R (2006) MTA1, a transcriptional activator of breast cancer amplified sequence 3. Proc Natl Acad Sci USA 103:6670–6675

Inoue Y, Itoh Y, Abe K, Okamoto T, Daitoku H, Fukamizu A, Onozaki K, Hayashi H (2007)Smad3 is acetylated by p300/CBP to regulate its transactivation activity. Oncogene 26:500–508

Ishinaga H, Jono H, Lim JH, Kweon SM, Xu H, Ha UH, Xu H, Koga T, Yan C, Feng XH, Chen LF, Li JD (2007) TGF-beta induces p65 acetylation to enhance bacteria-induced NF-kappaB activation. Embo J 26:1150–1162

Ito K, Yamamura S, Essilfie-Quaye S, Cosi B, Ito M, Barnes PJ, Adcock IM (2006) Histone deacetylase 2-mediated deacetylation of the glucocorticoid receptor enables NF-kappaB suppression. J Exp Med 203:7–13

Jeong JW, Bae MK, Ahn MY, Kim SH, Sohn TK, Bae MH, Yoo MA, Song EJ, Lee Kj, Kim KW, (2002) Regulation and destabilization of HIF-1alpha by ARD1-mediated acetylation. Cell 111:709–720

Kleff S, Andrulis ED, Anderson CW, Sternglanz (1995) Identification of a gene encoding a yeast histone H4 acetyltransferase. J Biol Chem 270:24674–24677

Knights CD, Catania J, Giovanni SD, Muratoglu S, Perez R, Awartzbeck A, Quong AA, Zhang X, Beerman T, Pestell RG, Avantaggiati ML (2006) Distinct p53 acetylation cassettes differentially influence gene-expression patterns and cell fate. J Cell Biol 173:533–544

Knudson KE, Cavenee WK, Arden KC (1999) D-type cyclins complex with the androgen receptor and inhibit its transcriptional transactivation ability. Cancer Res 59:2297–2301

Lin HY, Hopkins R, Cao HJ, Tang HY, Alexander C, Davis FB, Davis PJ (2005) Acetylation of nuclear hormone receptor superfamily members: thyroid hormone causes acetylation of its own receptor by a mitogen-activated protein kinase-dependent mechanism. Steroids 70:444–449

Liu MC, Marshall JL, Pestell RG (2004) Novel strategies in cancer therapeutics: targeting enzymes involved in cell cycle regulation and cellular proliferation. Curr Cancer Drug Targets 4:403–424

Mittal R, Peak-Chew SY, McMahon HT (2006) Acetylation of MEK2 and I kappa B kinase (IKK) activation loop residues by YopJ inhibits signaling. Proc Natl Acad Sci USA 103:18574–18579

Pasheva E, Sarov M, Bidjekov K, Ugrinova I, Sarg B, Lindner H, Pashev IG (2004) In vitro acetylation of HMGB-1 and -2 proteins by CBP: the role of the acidic tail. Biochemistry 43:2935–2940

Pestell RG, Jameson JK (1995) Hormone action II: transcriptional regulation of endocrine genes by second messenger signalling pathways. Chapter 5. In: Weintraub B (ed) Molecular endocrinology: basic concepts and clinical correlations. New York: Raven Press pp 59–76

Pestell RG, Albanese C, Reutens AT, Segall JE, Lee RJ, Arnold A (1999) The cyclins and cyclin-dependent kinase inhibitors in hormonal regulation of proliferation and differentiation. Endocrine Rev 20:501–534

Reutens AT, Fu M, Wang C, Alvanese C, McPhaul MJ, Sun Z, Balk SP, Janne OA, Palvimo JJ, Restell RG (2001) Cyclin D1 binds the androgen receptor and regulates hormone-dependent signaling in a p300/CBP-associated factor (P/CAF)-dependent manner. Mol Endocrinol 15:797–811

Sakamaki R, Casimiro MC, Ju X, Quong AA, Katiyar S, Liu M, Jiao X, Li A, Zhang X, Lu Y, Wang C, Byers S, Nicholson R, Link T, Shemluck M, Yang J, Fricke ST, Novikoff PM, Papanikolaou A, Arnold A, Albanese C, Pestell R (2006) Cyclin D1 determines mitochondrial function in vivo. Mol Cell Biol 26:5449–5469

Scroggins BT, Robzyk K, Wang D, Marcu MG, Trutsumi S, Beebe K, Cotter RJ, Felts S, Toft D, Karnitz L, Rosen N, Neckers L (2007) An acetylation site in the middle domain of Hsp90 regulates chaperone function. Mol Cell 25:151–159

Simonsson M, Kanduri M, Gronroos E, Heldin CH, Ericsson J (2006) The DNA binding activities of Smad2 and Smad3 are regulated by coactivator-mediated acetylation. J Biol Chem 281:39870–39880

Sridhar J, Pattabiraman N, Pestell RG (2005) CDK inhibitors as anticancer gents. Chapter 27. In: Yue E, Smith PJ (eds) Inhibitors of cyclin-dependent kinases as anti-tumor agents. Vol. 2. CRC Press, Boca Raton, Fl

Strahl BD, Allis CD (2000) The language of covalent histone modifications. Nature 403:41–45

Wang C, Fu M, Angeletti RH, Siconolgi-Baez L, Reutens AT, Albanese C, Lisanti MP, Katzenellenbogen BS, Kato S, Hopp T, Fuqua SA, Lopesz GN, Kushner PJ, Pestell RG (2001) Direct acetylation of the estrogen receptor alpha hinge region by p300 regulates transactivation and hormone sensitivity. J Biol Chem 276:18375–18383

Wang C, Pattabiraman N, Zhou JN, Fu M, Salamaki T, Albanese C, Li Z, Wu K, Hulit J, Neumeister P, Novikoff PM, Brownlee M, Scherer PE, Jones JG, Whitney KD, Donehower LA, Harris EL, Rohan T, Johns DC, Pestell RG (2003) Cyclin D1 repression of peroxisome proliferator-activated receptor gamma expression and transactivation. Mol Cell Biol 23:6159–6173

Wnag C, Li Z, Lu Y, Du R, Katiyar S, Yang J, Fu M, Leader JE, Quong A, Novikoff PM, Pestell RG (2006) Cyclin D1 repression of nuclear respiratory factor 1 integrates nuclear DNA synthesis and mitochondrial function. Proc Natl Acad Sci USA 103:11567–11572

Wolf D, Rodova M, Miska EA, Calvet JP, Kouzarides T (2002) Acetylation of beta-catenin by CREB-binding protein (CBP). J Biol Chem 277:25562–25567

A Novel Function for Cyclin E in Cell Cycle Progression

Yan Geng[1], *Youngmi Lee*[1], *Markus Welcker*[2], *Jherek Swanger*[2], *Agnieszka Zagozdzon*[1], *James M. Roberts*[3], *Philipp Kaldis*[4], *Bruce E. Clurman*[2], and *Piotr Sicinski*[1]

Summary

In this study we demonstrated the presence of a kinase-independent function for cyclin E. Specifically, we observed that a kinase-deficient cyclin E1 mutant can reconstitute cyclin E's function in cyclin E-null cells. Kinase-deficient cyclin E1 is loaded onto chromatin during $G_0 \rightarrow$ S progression, it restores MCM incorporation and it facilitates S phase entry of cyclin E-null cells. We also observed that, in wild-type cells, cyclin E is loaded onto DNA during the $G_0 \rightarrow$ S transition, and it co-localizes with MCM on chromatin. We demonstrated a physical interaction between cyclin E and MCM. We propose that the DNA-bound fraction of cyclin E facilitates MCM loading in a kinase-independent fashion. Our work indicates that, in addition to their well-established function as activators of cyclin-dependent kinases, E-cyclins play a kinase-independent function in cell cycle progression.

Results and Discussion

Cyclins E1 and E2 (E-type cyclins) are components of the core cell cycle machinery in mammalian cells. E-cyclins are thought to function as regulatory, activating subunits of cyclin-dependent kinases (CDKs), primarily CDK2 (Dulic et al. 1992; Koff et al. 1992). Cyclin E-CDK2 complexes phosphorylate several cellular proteins, thereby contributing to cell cycle progression (Hwang and Clurman 2005).

We and others previously generated cyclin E-null mice using gene targeting. Cyclin $E1^{-/-}E2^{-/-}$ animals died at embryonal day E11.5 due to abnormal placental development (Geng et al. 2003; Parisi et al. 2003). Fibroblasts isolated from cyclin E-null embryos displayed a defect in $G_0 \rightarrow$ S phase progression (Geng et al. 2003; Parisi et al. 2003). In contrast, mice lacking CDK2 – the major catalytic partner of cyclin E – were viable and developed relatively normally, with the exception of gonads (Berthet et al. 2003; Ortega et al. 2003). Moreover, the proliferation of CDK2-null fibroblasts was relatively unperturbed, and $CDK2^{-/-}$ cells displayed normal $G_0 \rightarrow$ S progression (Berthet et al. 2003; Ortega et al. 2003). These observations raised the possibility that some functions of the E-cyclins might be CDK-independent.

[1] Department of Cancer Biology, Dana-Farber Cancer Institute and Department of Pathology, Harvard Medical School, Boston, Massachusetts 02115, USA, *e-mail*: peter_sicinski@dfci.harvard.edu

[2] Divisions of Clinical Research and Human Biology

[3] Howard Hughes Medical Institute, Division of Basic Sciences, Fred Hutchinson Cancer Research Center, Seattle, Washington 98109, USA

[4] Mouse Cancer Genetics Program, National Cancer Institute, Frederick, Maryland 21702, USA

Melmed et al.
Hormonal Control of Cell Cycle
© Springer-Verlag Berlin Heidelberg 2008

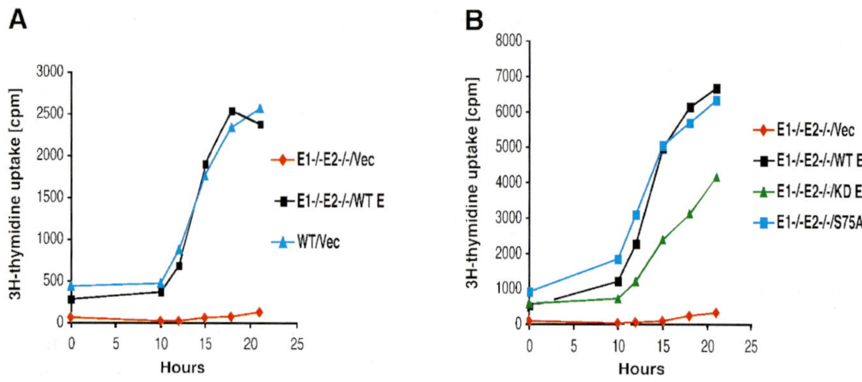

Fig. 1. Rescue of cyclin E-null phenotypes by kinase-deficient cyclin E. Wild-type (WT) or cyclin E-null (E1$^{-/-}$E2$^{-/-}$) MEFs were transduced with an empty vector (Vec), or with retroviruses encoding wild-type human cyclin E1 (WT E), cyclin E1 S75A mutant (S75A), or with kinase-deficient cyclin E1 EEIYP mutant (KD E). Cells were serum starved and then forced to re-enter the cell cycle by serum addition. Incorporation of [^{3}H]-thymidine was determined at the indicated time points

In the current study, we investigated this possibility using cyclin E-null cells. We and others had previously shown that the E-cyclins were required for the re-entry of cells from quiescence, because cyclin E1$^{-/-}$E2$^{-/-}$ mouse embryo fibroblasts (MEFs) were unable to re-enter the cell cycle from the quiescent, G_0 state (Geng et al. 2003; Parisi et al. 2003). We then asked which function of cyclin E underlies the critical requirement for this protein in $G_0 \rightarrow$ S progression. To address this question, we transduced cyclin E1$^{-/-}$E2$^{-/-}$ MEFs (further referred to as cyclin E-null cells) with retroviruses encoding human wild-type cyclin E1 or with various cyclin E1 mutants. We verified that the levels of the ectopically expressed human cyclin E1 matched the levels of endogenous mouse cyclin E1 in wild-type cells. We chose to re-constitute cyclin E1 (rather then both E-type cyclins) because cells expressing a single E-type cyclin (E1$^{-/-}$ cells or E2$^{-/-}$ cells) were previously shown to behave like their wild-type counterparts (Geng et al. 2003; Parisi et al. 2003). Consistent with our previous findings, cyclin E-null cells that were transduced with empty vectors failed to re-enter the S phase from quiescence (Fig. 1A). As expected, the introduction of wild-type cyclin E1 into cyclin E-null cells fully restored the ability of these cells to re-enter the S-phase (Fig. 1A). Very surprisingly, we found that the cyclin E1 kinase-deficient mutant was able to support entry of cells into the S-phase, albeit at a reduced rate as compared to wild-type cyclin E1 (Fig. 1B). The kinase-deficient cyclin E1 is an alanine scanning mutant, in which residues 188–192 (EEIYP) were changed to alanines (Sheaff et al. 1997). A recent crystal structure revealed that these cyclin E residues make direct contacts with the CDK2 activation segment and CDK2 pT160, regions of CDK2 that are critical for substrate recognition (Honda et al. 2005). The EEIYP mutant was previously shown to be unable to direct phosphorylation of histone H1 (Sheaff et al. 1997), and it was unable to phosphorylate another cyclin E-CDK target, p27^{Kip1}, as evidenced by the inability of this cyclin E1 mutant to overcome cell cycle block induced by p27^{Kip1} overexpression (Sheaff et al. 1997).

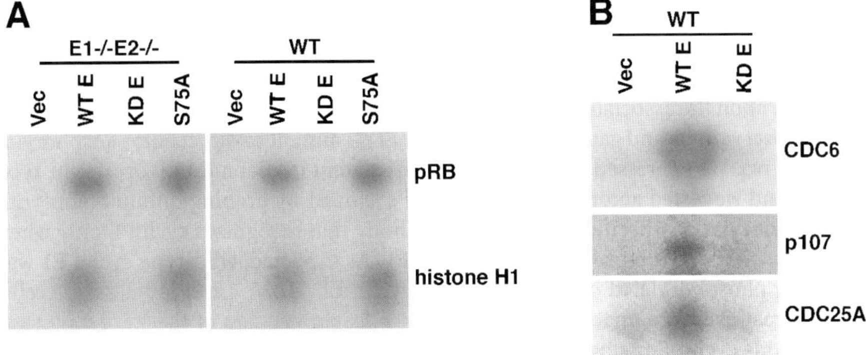

Fig. 2. Analyses of kinase-deficient EEIYP cyclin E1 mutant. Wild-type (WT) or cyclin E-null $(E1^{-/-}E2^{-/-})$ MEFs were infected with empty vector (Vec), or with retroviruses encoding wild-type (WT E), EEIYP mutant (KD E) or S75A mutant (S75A) cyclin E1. Ectopically expressed cyclin E was immunoprecipitated, and in vitro kinase reaction was performed using the indicated proteins as substrates

Immunoprecipitation of EEIYP cyclin E1 from wild-type and cyclin $E1^{-/-}E2^{-/-}$ MEFs revealed that this mutant retained the ability to interact with CDK2. Although cyclin E was shown to interact with CDK1 in certain settings (Aleem et al. 2005), we found that neither wild-type nor the EEIYP mutant bound to CDK1 in these cells. As expected, cyclin E1 did not associate with CDK4 in wild-type or cyclin E-null cells.

We next immunoprecipitated wild-type cyclin E1 or the EEIYP mutant from wild-type or from cyclin E-null MEFs and performed in vitro kinase reactions using histone H1 or retinoblastoma protein (pRB) as a substrate. Since the cyclin E mutants used in our studies were of human origin, we were able to bring down ectopically expressed proteins, but not endogenous mouse cyclin E, using an antibody against human cyclin E. We found that the EEIYP mutant was unable to phosphorylate either pRB or histone H1 proteins (Fig. 2A).

We next tested the ability of immunoprecipitated EEIYP mutant to direct phosphorylation of other known cyclin E-CDK targets, namely CDC6, CDC25 and p107. Again, we found that this mutant was unable to phosphorylate all these substrates (Fig. 2B).

The kinase activity of cyclin E-CDK complexes is normally restrained by cell cycle inhibitors $p27^{Kip1}$ and $p21^{Cip1}$. Consequently, activity of cyclin E-CDK kinase is greatly increased in cells lacking these proteins (Aleem et al. 2005). To test the kinase activity of the EEIYP mutant under this condition, we expressed cyclin E1 EEIYP mutant in MEFs derived from $p27^{Kip1-/-}p21^{Cip1-/-}$ mice. We immunoprecipitated the mutant and performed in vitro kinase reactions using histone H1 as a substrate. Again, we found that the cyclin E1 mutant was unable to phosphorylate this substrate. Hence, cyclin E1 EEIYP mutant is deficient in activating the associated kinase even in $p27^{Kip1}/p21^{Cip1}$-null cells, i.e., under conditions where wild-type cyclin E-associated kinase activity is hyperactivated.

To rigorously test the ability of the EEIYP mutant to direct the phosphorylation of cyclin E-CDK substrates in vivo, we searched for a substrate that could be phosphorylated exclusively by cyclin E-CDK but not by other kinases. To the best of our knowledge,

A

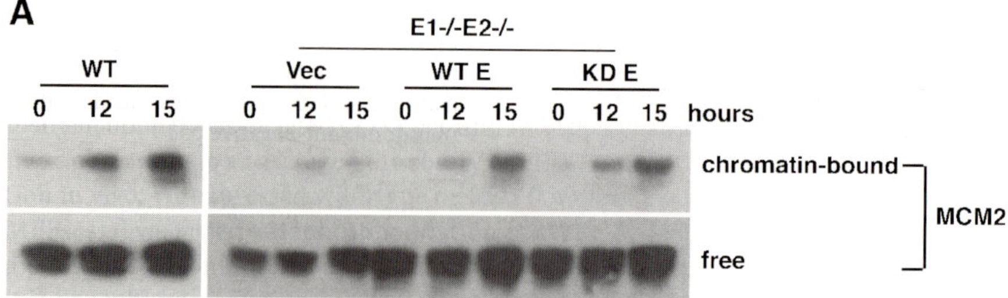

B

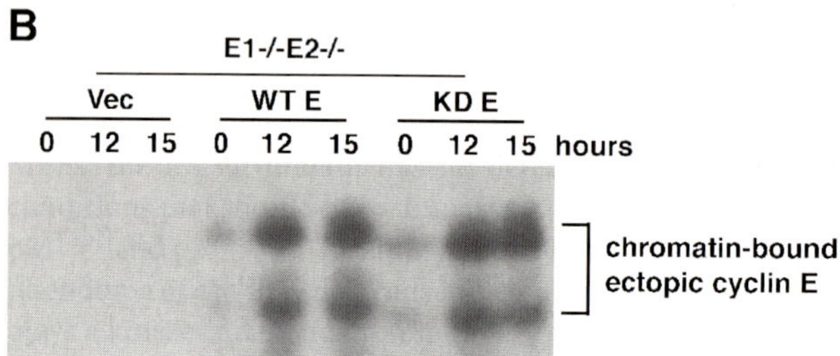

C

D

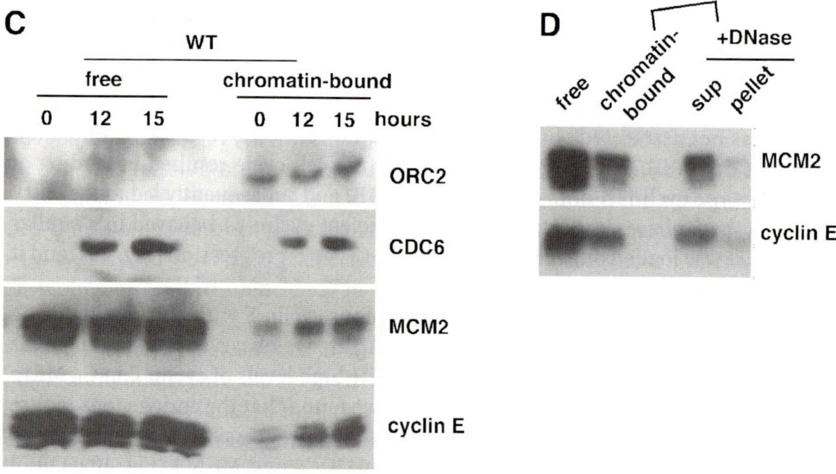

Fig. 4. Loading of MCM and cyclin E onto chromatin during cell cycle re-entry of wild-type (WT) and cyclin E-deficient (E1$^{-/-}$E2$^{-/-}$) MEFs. (**A**) Chromatin-bound MCM2 and its levels in the "free" fraction were determined by Western blotting. (**B**) Chromatin-bound human cyclin E ectopically expressed in cyclin E1$^{-/-}$E2$^{-/-}$ MEFs. (**C**) Chromatin-bound endogenous cyclin E1, CDC6, MCM2 and ORC2 and their levels in the "free" fraction were determined by Western blotting. (**D**) Chromatin-bound fraction was digested with DNase I, and the presence of cyclin E and MCM2 in the supernatant (sup) and pellet are shown

A

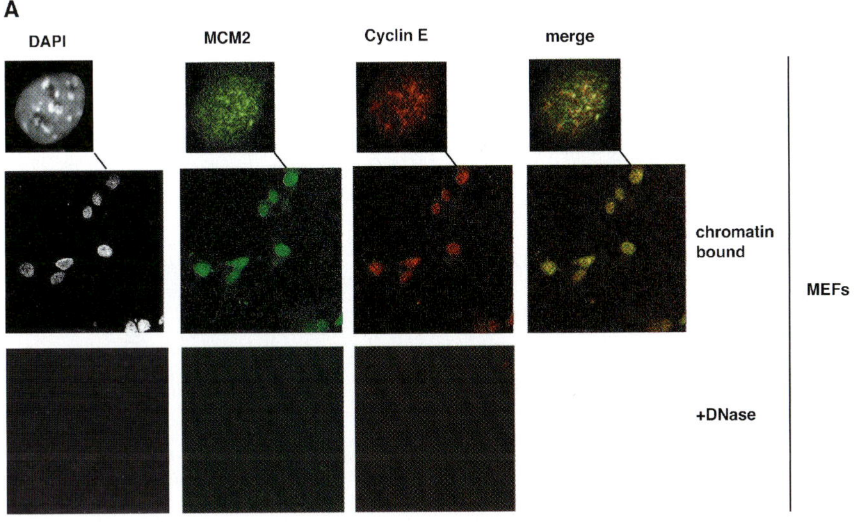

B

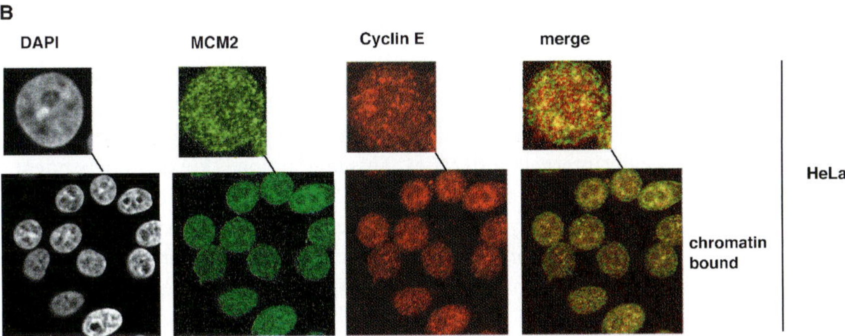

Fig. 5. Cyclin E and MCM are present in the chromatin-bound fraction of MEFs (**A**) and human Hela cells (**B**)

cells revealed the presence of cyclin E in the chromatin-bound fraction (Fig. 5A,B). As expected, MCM2 was also found in this fraction (Fig. 5A,B). Treatment of the extracted cells with DNase released cyclin E and MCM from cells, confirming that these proteins were DNA-bound (Fig. 5A).

Lastly, we asked whether cyclin E might physically associate with MCMs. We ectopically expressed cyclin E in H293 cells and immunoprecipitated cyclin E followed by immunoblotting with antibodies against MCM2 and MCM7. These analyses revealed that cyclin E co-immunoprecipitated with endogenous MCMs (Fig. 6A). We also verified cyclin E-MCM interaction by incubating GST-MCM2 or GST-MCM7 fusion proteins with H293 cell lysates, followed by immunoblotting with anti-cyclin E anti-

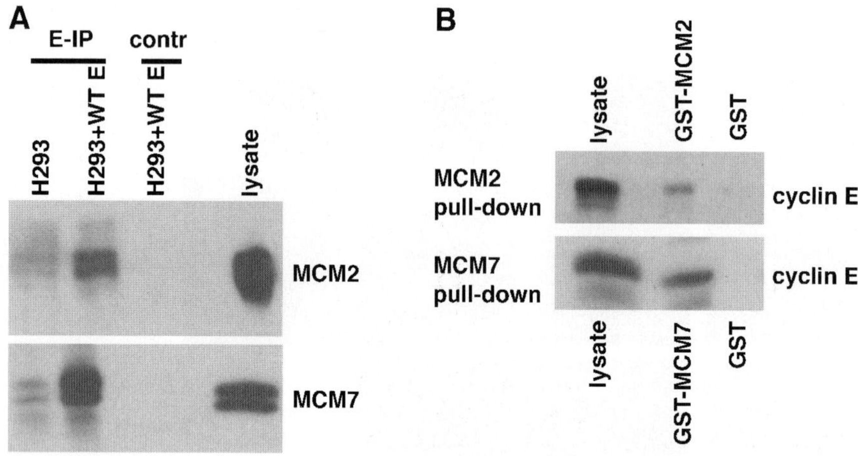

Fig. 6. Physical interaction of cyclin E with MCM2 and MCM7. (**A**) Co-immunoprecipitation of cyclin E1 with endogenous MCMs. (**B**) Pull-down of cyclin E1 in H293 cells by GST-MCM2 or GST-MCM7 proteins

bodies. Again, we observed interaction of MCMs with endogenous cyclin E (Fig. 6B). We also observed that in vitro translated cyclin E bound to GST-MCM2 and to GST-MCM7. Hence, all these approaches confirmed that cyclin E can physically associate with MCMs.

The results described here reveal that, in addition to their very well-established functions as activators of CDKs, E cyclins play an additional kinase-independent function in cell cycle re-entry. We have shown that a fraction of cyclin E is localized on chromatin, where it physically interacts with MCM. This interaction is kinase-independent, and the kinase-deficient cyclin E mutant restored the ability of cyclin E-null cells to load MCM and to re-enter cell cycle. These results suggest that cyclin E facilitates MCM loading in a kinase-independent fashion. Consistent with this thinking, others have shown that incorporation of MCM into DNA replication origins proceeds in mammalian cells even when CDK activity is inhibited by p21^{Cip1} expression (Cook et al. 2002) or by treatment of cells with butyrolactone I (which inhibits CDK2 and CDK1; Arata et al. 2000). Cyclin E was shown to be loaded onto chromatin in Xenopus extracts (Chevalier et al. 1996; Furstenthal et al. 2001), where it interacts with DNA-bound CDC6 (Furstenthal et al. 2001). This loading of cyclin E did not require CDK kinase activity, as addition of the chemical CDK inhibitor, roscovitine, had no effect on cyclin E chromatin recruitment (Furstenthal et al. 2001). These observations raise the possibility that cyclin E might form a physical "bridge" between components of the pre-replication complexes and MCMs, thereby contributing to MCM loading.

Cyclin E overexpression is involved in many human cancers (Donnellan and Chetty 1999). The major catalytic partner of cyclin E, CDK2, was shown to be dispensable for proliferation of several cancer cell lines (Tetsu and McCormick 2003). It remains to be seen whether the kinase-independent function of cyclin E, identified in this study, is required for cancer cell proliferation. It is tempting to speculate that this function is responsible, at least in part, for the oncogenic action of cyclin E. The kinase-independent function of the overexpressed cyclin E might contribute to the escape of tumor cells from quiescence, thereby contributing to cancer formation.

References

Aleem E, Kiyokawa H, Kaldis P (2005) Cdc2-cyclin E complexes regulate the G1/S phase transition. Nature Cell Biol 7:831–836

Arata Y, Fujita M, Ohtani K, Kijima S, Kato JY (2000) Cdk2-dependent and -independent pathways in E2F-mediated S phase induction. J Biol Chem 275:6337–6345

Berthet C, Aleem E, Coppola V, Tessarollo L, Kaldis P (2003) Cdk2 knockout mice are viable. Curr Biol 13:1775–1785

Chevalier S, Couturier A, Chartrain I, Le Guellec R, Beckhelling C, Le Guellec K, Philippe M, Ford CC (1996) Xenopus cyclin E, a nuclear phosphoprotein, accumulates when oocytes gain the ability to initiate DNA replication. J Cell Sci 109 (Pt 6):1173–1184

Cook JG, Park CH, Burke TW, Leone G, DeGregori J, Engel A, Nevins JR (2002) Analysis of Cdc6 function in the assembly of mammalian prereplication complexes. Proc Natl Acad Sci USA 99:1347–1352

Coverley D, Laman H, Laskey RA (2002) Distinct roles for cyclins E and A during DNA replication complex assembly and activation. Nature Cell Biol 4:523–528

Donnellan R, Chetty R (1999) Cyclin E in human cancers. Faseb J 13:773–780

Dulic V, Lees E, Reed SI (1992) Association of human cyclin E with a periodic G1-S phase protein kinase. Science 257:1958–1961

Furstenthal L, Kaiser BK, Swanson C, Jackson PK (2001) Cyclin E uses Cdc6 as a chromatin-associated receptor required for DNA replication. J Cell Biol 152:1267–1278

Geng Y, Yu Q, Sicinska E, Das M, Schneider JE, Bhattacharya S, Rideout WM, Bronson RT, Gardner H, Sicinski P (2003) Cyclin E ablation in the mouse. Cell 114:431–443

Honda R, Lowe ED, Dubinina E, Skamnaki V, Cook A, Brown NR, Johnson LN (2005) The structure of cyclin E1/CDK2: implications for CDK2 activation and CDK2-independent roles. Embo J 24:452–463

Hwang HC, Clurman BE (2005) Cyclin E in normal and neoplastic cell cycles. Oncogene 24:2776–2786

Koff A, Giordano A, Desai D, Yamashita K, Harper JW, Elledge S, Nishimoto T, Morgan DO, Franza BR, Roberts JM (1992) Formation and activation of a cyclin E-cdk2 complex during the G1 phase of the human cell cycle. Science 257:1689–1694

Madine MA, Swietlik M, Pelizon C, Romanowski P, Mills AD, Laskey RA (2000) The roles of the MCM, ORC, and Cdc6 proteins in determining the replication competence of chromatin in quiescent cells. J Struct Biol 129:198–210

Martini E, Roche DM, Marheineke K, Verreault A, Almouzni G (1998) Recruitment of phosphorylated chromatin assembly factor 1 to chromatin after UV irradiation of human cells. J Cell Biol 143:563–575

Ortega S, Prieto I, Odajima J, Martin A, Dubus P, Sotillo R, Barbero JL, Malumbres M, Barbacid M (2003) Cyclin-dependent kinase 2 is essential for meiosis but not for mitotic cell division in mice. Nature Genet 35:25–31

Parisi T, Beck AR, Rougier N, McNeil T, Lucian L, Werb Z, Amati B (2003) Cyclins E1 and E2 are required for endoreplication in placental trophoblast giant cells. Embo J 22:4794–4803

Sheaff RJ, Groudine M, Gordon M, Roberts JM, Clurman BE (1997) Cyclin E-CDK2 is a regulator of p27Kip1. Genes Dev 11:1464–1478

Tetsu O, McCormick F (2003) Proliferation of cancer cells despite CDK2 inhibition. Cancer Cell 3:233–245

Welcker M, Singer J, Loeb KR, Grim J, Bloecher A, Gurien-West M, Clurman BE, Roberts JM (2003) Multisite phosphorylation by Cdk2 and GSK3 controls cyclin E degradation. Mol Cell 12:381–392

IGF-I and the Regulation of Cell Cycle Progression in Smooth Muscle Cells

David R. Clemmons[1]

Summary

Insulin-like growth factor-I is a mitogen for multiple cell types. Vascular smooth muscle cells (SMC) are a useful model for studying cell cycle regulation because they maintain a stable, partially dedifferentiated phenotype in vitro and their phenotypic characteristics are similar to the characteristics of vascular SMC that proliferate in whole animal models following vascular injury. Exposure of cultured SMC to serum results in a doubling within 48 hr. In contrast, these cells require the induction of cell stress to proliferate in response to IGF-I. Hyperglycemia serves as a model of cell stress in this test system. Cells exposed to 25 mM glucose will double in number in response to IGF-I within 48 hr whereas cells exposed to 5 mM glucose do not proliferate. There are two important determinants of the ability of these cells to respond to IGF-I with an increase in proliferation. The first is that the signaling protein, Shc, must be recruited to the cell membrane and phosphorylated. Induction of Shc phosphorylation results in enhancement of MAP kinase activity, which is required for an increase in cell division. In order for these cells to undergo a mitogenic response and for Shc to be phosphorylated, the signaling protein IRS-1 must be down regulated, which occurs in these cells in response to hyperglycemic stress. Second, to recruit Shc to the cell membrane, the integrin receptor, $\alpha V\beta 3$, must be stimulated by increasing concentrations of its ligands. Hyperglycemic stress induces the increased synthesis of several $\alpha V\beta 3$ ligands, including vitronectin, osteopontin and thrombospondin. Blocking the binding of these ligands to $\alpha V\beta 3$ results in an inability to induce Shc in response to IGF-I and failure of the cells to proliferate. The changes that are induced by hyperglycemia that result in enhanced IGF-I responsiveness are accompanied by changes in cell cycle marker proteins. IGF-I induces cyclin E and CDK_2 as well as cyclin D1 and CDK_4. These changes are not mediated by changes in PI-3 kinase or AKT activation but do require the activation of mTOR, since they can be inhibited by rapamycin. In summary, hyperglycemic stress induces a change in $\alpha V\beta 3$ ligand occupancy, which then signals cooperatively with the IGF-I receptor to activate Shc induction. These changes lead to activation of the MAP kinase pathway, which results in an increase in the cell proliferation in response to IGF-I. Understanding the mechanism by which hyperglycemia sensitizes cells to the growth

---

[1] Department of Medicine, UNC School of Medicine, Chapel Hill, NC 27599 USA

Address to correspondence: CB# 7170, 8024 Burnett-Womack, Division of Endocrinology, University of North Carolina at Chapel Hill, Chapel Hill, NC 27599-7170

e-mail: endo@med.unc.edu

Melmed et al.
Hormonal Control of Cell Cycle
© Springer-Verlag Berlin Heidelberg 2008

stimulatory effects of IGF-I is likely to open up new opportunities for therapeutic intervention.

Introduction

Insulin-like growth factor-I (IGF-I) is a potent mitogen for multiple mesenchymal cell types. Vascular smooth muscle cells (SMC) have abundant IGF-I receptors and respond to IGF-I with increases in both cell migration and proliferation, which are necessary to form atherosclerotic lesions within blood vessels (Margariti et al. 2006). In vivo models have supported a role for IGF-I in the proliferative phase of atherosclerosis. Specifically, following vascular injury by balloon denudation, there is an increase in IGF-I synthesis that peaks at 7–10 days, the time when there is maximal increase in cell proliferation (Khorsandi et al. 1992; Cercek et al. 1990). Various methods have been utilized to inhibit the effect of this locally secreted IGF-I on adjacent SMC, including infusion of anti IGF-I receptor antibodies, administration of a peptide that blocks IGF-I binding to its receptor, and over-expression of a protease-resistant form of insulin-like growth factor binding protein that inhibits IGF-I binding to its receptors. Results from studies utilizing all three of these methodologies have suggested that blocking IGF-I binding to its receptor results in attenuation of SMC proliferation (Zhu et al. 2001; Zhang et al. 2002; Hayry et al. 1995).

Like most anchorage-dependent cells, vascular SMC proliferate when anchored to an ECM, but if they are analyzed in suspension, their proliferative response to mitogens is markedly reduced. Based on this observation, our laboratory has been interested in cooperative signaling between integrin receptors, which mediate cell attachment, and IGF-I receptor-linked signaling mechanisms. Studies in other cell types have shown that there is cooperative signaling between either VEGF receptors or PDGF receptors and integrins (Miyamoto et al. 1996; Senger et al. 1997). Initially we demonstrated that blocking ligand occupancy of the $\alpha V \beta 3$ integrin resulted in attenuation of IGF-I actions (Jones et al. 1996). $\alpha V \beta 3$ is a heterodimer composed at one α and one β subunit that is abundantly expressed on vascular endothelium and SMC as well as on osteoclasts (Horton 1997). In contrast to these three cell types, most other cell types in normal mammalian systems have very few $\alpha V \beta 3$ receptors. Therefore, the mechanisms that link signaling between this integrin and the IGF-I receptor are likely to be limited to a few cell types. We were able to demonstrate in vitro that addition of the disintegrin, echistatin, which is a competitive inhibitor of ligand binding, to $\alpha V \beta 3$ resulted in attenuation of the ability of IGF-I to stimulate SMC migration or division (Jones et al. 1996; Zheng and Clemmons 1998). Further studies demonstrated that, in a porcine model of atherosclerosis, when $\alpha V \beta 3$ antagonists that were similar in structure to echistatin were infused into pig blood vessels following injury, there was marked attenuation of atherosclerotic lesion development; specifically lesion size was reduced by 48% (Nichols et al. 1999). To determine that this was related to altered IGF-I actions, a marker protein, insulin-like growth factor binding protein-5, whose transcription is directly increased by IGF-I stimulation, was quantified. Following exposure to $\alpha V \beta 3$ inhibitors, there was a 5-fold reduction in IGFBP-5 expression by lesion tissue whereas control vessels that received a dummy compound had no change in IGFBP-5 expression. These findings clearly indicated that there was cooperative signaling between IGF-I and the αVβ3 integrin in vivo.

Determination of the Molecular Mechanism by Which the β3 and the IGF-I Receptor Cooperatively Signal to Stimulate SMC Division

To define the mechanism that mediates cooperative signaling, we initially analyzed the effects of increasing ligand occupancy of αVβ3. Our studies determined that addition of ligands resulted in phosphorylation of two critical tyrosines within two NPXY sequences in the β3 cytoplasmic tail (Ling et al. 2003). We determined that once these tyrosines were phosphorylated, they served as docking sites for a protein termed DOK-1 (Ling et al. 2005a). Furthermore exposure of cells to IGF-I resulted in DOK-1 phosphorylation which resulted in its binding to an intracellular phosphatase termed SHP-2. While SHP-2 cannot directly bind to β3, its localization on DOK-1 resulted in its recruitment from the cytoskeleton to the cell membrane. We further determined that blocking ligand occupancy of β3 in SMC resulted in failure of IGF-I to be able to stimulate SHP-2 localization to the cell membrane. Our next series of studies determined that, once SHP-2 is localized to the cell membrane following IGF-I stimulation, it was transferred to a scaffold protein termed SHPS-1 (Oshima et al. 2002). SHPS-1 contains 4 tyrosines contained within YXXL/I/V motifs within its cytoplasmic domain. These tyrosines are phosphorylated in response to IGF-I receptor stimulation. Following their phosphorylation, they act as important docking sites for SHP-2. However, as stated previously, SHP-2 must be localized in the plasma membrane in order to be transferred to phosphorylated SHPS-1. Blocking SHP-2 transfer to SHPS-1 blocked the ability of IGF-I to stimulate SMC migration and/or division; therefore, this is considered a critical event for signaling (Maile and Clemmons 2002).

To determine how this event related to the ability of IGF-I to stimulate a mitogenic response, we subsequently determined that it was critical to activate the MAP kinase pathway in order for IGF-I to stimulate mitogenesis (Imai and Clemmons 1999). Furthermore, unlike in most cell types, in vascular SMC the MAP kinase pathway was not activated in response to activation of the signaling protein IRS-1 but rather it required activation of Shc. We prepared a Shc mutant that had its three critical tyrosines changed to phenylalanines and could not bind Grb2. Expression of this mutant resulted in the complete inability of IGF-I to stimulate SMC migration or division. Therefore, Shc phosphorylation and localization of Grb2 are required to activate the Ras/MAP kinase system in these cells and thus are required for stimulation of cell division (Ling et al. 2005b). This finding raised the important question as to how SHP-2 localization on SHPS-1 resulted in Shc phosphorylation. In subsequent studies we determined that, following β3 ligand occupancy and IGF-I receptor phosphorylation, there was a marked increase in the activation of Src kinase and that Src kinase bound directly to SHP-2 through the interaction of a polyproline sequence within SHP-2 and a Src SH3 domain. This resulted in autoactivation of Src and phosphorylation of tyrosine 295 in the activation loop. Activated Src then phosphorylated an additional c-Src tyrosine that was required for Shc binding to Src. Once Shc was bound to c-Src, it could be directly phosphorylated by the Src kinase activity and this activation occurred on SHPS-1.

(Lieskovska et al. 2006). Therefore, the recruitment of a ternary complex containing SHP-2, Src and Shc to SHPS-1 results in Shc phosphorylation and subsequent activation of downstream signaling. Utilization of mutagenesis or competitive inhibitory peptides as well as siRNA showed that disruption of either the SHP-2/Src complex or the Src/Shc complex resulted in failure to initiate Shc phosphorylation and inhibition of the ability of IGF-I to stimulate cell proliferation. These results further proved that cooperative signaling between the $\alpha V\beta 3$ integrin receptor and the IGF-I receptor is required for SMC to proliferate in response to IGF-I, and they provide a mechanism for understanding why the mitogenic response requires that both components of this signaling system are activated.

Effect of Hyperglycemia on IGF-I Signaling and SMC Proliferation

Hyperglycemia has been shown to induce cell stress. Specifically, various studies have shown that when cells are exposed to high glucose they increase their production of reactive oxygen species, in addition, induction of endoplasmic reticulum (ER)stress is thought to be an important component of the adaptive response (Haidara et al. 2006; Ozawa et al. 2005). These changes have been shown to lead to two important changes that may impact IGF-IR/$\alpha V\beta 3$-linked signaling. First, high glucose has been shown to induce phosphorylation of Jnk2, which phosphorylates serine 307 on IRS-1, resulting in its targeting to a proteosome and leading to down regulation of IRS-1 (Ozcan et al. 2006; Hiratani et al. 2005; Izawa et al. 2005). Second, hyperglycemia has been shown to lead to a marked increase in the synthesis and secretion of $\alpha V\beta 3$ ligands, specifically thrombospondin, osteopontin and vitronectin (Kawamura et al. 2004; Holmes et al. 1997; Stenina et al. 2003). To confirm that these changes occurred in SMC in response to hyperglycemia, we undertook several experiments. Initially we were able to demonstrate that cells cultured in 25 mM glucose responded to IGF-I with a much greater proliferative response. The addition of IGF-I resulted in a two-fold increase in cell number in 48 hr whereas cells maintained in 5 mM glucose showed no increase in proliferation in response to IGF-I. Similarly, cell migration was preferentially stimulated in response to prolonged exposure of SMC to 25 mM glucose. Several control experiments were undertaken to rule out the possibility that these changes were due to a change in cell differentiation or due to inhibition of apoptosis rather than to an increase in proliferation.

Once these findings were established, we undertook studies to determine how the signaling mechanisms had changed. Initially we showed that cells cultured in 5 mM glucose had a major increase in IRS-1 phosphorylation in response to IGF-I. In contrast, cells cultured in 25 mM glucose had no change in IRS-1 tyrosine phosphorylation, and the levels of total IRS-1 protein were markedly reduced. That this was due to serine phosphorylation was demonstrated by showing that there was an excess of serine 307 phosphorylated IRS-1 in the cells exposed to 25 mM glucose, making it likely that this is the mechanism accounting for decreased IRS-1 concentrations. To further discern the cellular changes that occurred in response to hyperglycemia, we assessed Shc phosphorylation. Cells cultured in 25 mM glucose have a major induction of phosphorylated Shc whereas cells cultured in 5 mM glucose showed no increase in Shc phosphorylation. Since αVβ3 ligand occupancy was required to enhance Shc phosphorylation, we determined the effect of inhibiting ligand occupancy on Shc phosphorylation in 25 mM glucose. A specific antibody that reacted with a region of the β3 integrin, termed the C-loop binding domain, was shown to markedly inhibit ligand occupancy of β3 by vitronectin or osteopontin (Maile et al. 2006a). Inhibition resulted in the inability to induce Shc phosphorylation in response to IGF-I. To confirm that the change that occurred with hyperglycemia was related to ligand occupancy of β3, we utilized cells cultured in 5 mM glucose and added exogenous vitronectin or a peptide encompassing the region of vitronectin that bound to β3. Using either stimulus, we were able to demonstrate that cells cultured in 5 mM glucose Shc phosphorylation could be induced in response to IGF-I if they were preincubated with whole vitronectin or the vitronectin peptide (Maile et al. 2006b). Furthermore, preincubation with the C-loop antibody inhibited this response. To confirm that this inhibition was related to cell growth, we measured cell proliferation and MAP kinase induction. Again cells cultured in 5 mM glucose preincubated with vitronectin, or the vitronectin peptide, showed a marked increase in MAP kinase phosphorylation in response to IGF-I, and they had an increase in cell proliferation. These changes could be inhibited by simultaneous exposure to the anti β3 antibody. Therefore, these findings strongly support two conclusions: first that, when exposed to glycemic stress, cells down regulate IRS-1 and upregulate the production of αVβ3 ligands; and second, these events result in a major change in IGF-I signaling, resulting in the ability of IGF-I to phosphorylate Shc, which leads to MAP kinase activation and increased cell proliferation.

Induction of Cell Cycle-specific Markers by IGF-I in SMC

To confirm that these changes in response to hyperglycemia result in a proliferative response to IGF-I and that IGF-I is not merely inhibiting apoptosis, studies were conducted to determine the effect of changing glucose concentrations on the induction of specific cell cycle markers. When proliferating cell nuclear antigen phosphorylation (PCNA) levels were assessed, they were shown to be induced several fold by IGF-I in cells in 25 mM glucose after 12 hours; this induction was not detectable in cells exposed to 5 mM glucose. In contrast, when down regulation of p27 kip was examined, cells in both 5 mM and 25 mM glucose showed p27 KIP down regulation in response to IGF-I. Furthermore, this change was specifically inhibited by AKT inhibitors, indicating that it was linked to AKT activation. To further assess the effect of IGF-I on cell cycle marker proteins, we measured changes in cyclin D1, E, CDK_2 and CDK_4. In the presence of high glucose, there was major induction of cyclin D1 and CDK_4. Cyclin E was also induced although the changes in CDK_2 were of a lesser magnitude. To determine the effect of vitronectin addition, cells in 5 mM glucose were exposed to vitronectin plus IGF-I and the induction of cyclin D1 and CDK_2 was assessed. Addition of vitronectin to cells cultured in 5 mM glucose cells resulted in enhancement of the ability of IGF-I to stimulate cyclin D1 and, to some extent, enhancement of the CDK_2 response. To determine the signaling pathways that were required for induction of cyclin D1, CDK_4, and cyclin E, the effects of several inhibitors of IGF-I receptor linked signaling molecules were analyzed. PI-3 kinase and AKT inhibitors had no effect on the induction of these proteins. In contrast, inhibitors of MAP kinase or mTOR, specifically rapamycin, significantly inhibited the ability of IGF-I to induce cyclin D1, cyclin E and CDK$_4$ in the presence of hyperglycemia. Therefore, it appears that cell cycle complexes that are necessary for entry into DNA synthesis are activated in response to IGF-I in the presence of hyperglycemia and that this response requires both activation of mTOR and MAP kinase.

Conclusions

These studies have delineated the intracellular signaling mechanisms that are activated by IGF-I in SMC that are exposed to hyperglycemia. Normally, in vivo these cells are in a quiescent state and the induction of cell stress, either by mechanical injury or chemical exposure such as hyperglycemia, is required for them to respond with a proliferative response to IGF-I. The mechanism by which hyperglycemia induces this stress is an important focus for future studies and it is certainly possible it involves multiple pathways, including generation of reactive oxygen species, ER stress and/or longer term changes, such as nonenzymatic glycosylation of important cell surface signaling molecules. Future studies will be directed toward determining which of these changes occur in vivo in animal models of hyperglycemia and how they are related to the ability of SMC to leave the quiescent state and enter into a state wherein cell cycle marker proteins can be activated by IGF-I.

References

Cercek B, Fishbein MC, Forrester JS, Helfant RH, Fagin JA (1990) Induction of insulin-like growth factor-I messenger RNA in rat aorta after balloon denudation. Circ Res 66:1755–1760

Haidara MA, Yassin HZ, Rateb M, Ammar H, Zorkani MA (2006) Role of oxidative stress in development of cardiovascular complications in diabetes mellitus. Curr Vasc Pharmacol 4:215–227

Hayry P, Myllarniemi M, Aavik E, Alatalo S, Aho P, Yilmaz S, Raisanen-Sokolowski A, Cozzone G, Jameson BA, Baserga R (1995) Stablid D-peptide analog of insulin-like growth factor-1 inhibits smooth muscle cell proliferation after carotid ballooning injury in the rat. FASEB J 9:1336–1344

Hiratani K, Haruta T, Tani A, Kawahara J, Usui I, Kobayashi M (2005) Roles of mTOR and JNK in serine phosphorylation, translocation and degradation of IRS-1. Biochem Biophys Res Commun 335:836–842

Holmes DIR, Wahab NA, Mason RM (1997) Identification of glucose-regulated genes in human mesangial cells by mRNA differential display. Biochem and Biophys Res Comm 238:179–184

Horton, MA (1997) The alpha v beta 3 integrin "vitronectin receptor". Int J Biochem Cell Biol 29:721–725

Imai Y, Clemmons DR (1999) Roles of phosphatidylinositol 3-kinase and mitogen-activated protein kinase pathways in stimulation of vascular smooth muscle cell migration and deoxyriboncleic acid synthesis by insulin-like growth factor-I. Endocrinology 140:4228–4235

Izawa Y, Yoshizumi M, Fujita Y, Ali N, Kanematsu Y, Ishizawa K, Tsuchiya K, Obata T, Ebina Y, Tomita S, Tamaki T (2005) ERK1/2 activation by angiotensin II inhibits insulin-induced glucose uptake in vascular smooth muscle cells. Exp Cell Res 308:291–299

Jones JI, Prevette T, Gockerman A, Clemmons DR (1996) Ligand occupancy of the alpha-V-beta3 integrin is necessary for smooth muscle cells to migrate in response to insulin-like growth factor. Proc Natl Acad Sci USA 93:2482–2487

Kawamura H, Yokote K, Asaumi S, Kobayashi K, Fujimoto M, Maezwa Y, Saito Y, Mori S (2004) High glucose-induced upregulation of osteopontin is mediated via Rho/Rho kinase pathway in cultured rat aortic smooth muscle cells. Aterioscler Thromb Vas Biol 24:276–281

Khorsandi MJ, Fagin JA, Giannella-Neto D, Forrester JS, Cercek B (1992) Regulation of insulin-like growth factor-I and its receptor in rat aorta after balloon denudation. Evidence for local bioactivity. J Clin Invest 90:1926–1931

Lieskovska J, Ling Y, Badley-Clarke J, Clemmons DR (2006) The role of Src kinase in insulin-like growth factor-dependent mitogenic signaling in vascular smooth muscle cells. J Biol Chem 281:25041–25053

Ling Y, Maile LA, Clemmons DR (2003) Tyrosine phosphorylation of the beta3-subunit of the alphaVbeta3 integrin is required for membrane association of the tyrosine phosphatase SHP-2 and its further recruitment to the insulin-like growth factor I receptor. Mol Endocrinol 17:1824–1833

Ling Y, Maile LA, Badley-Clarke J, Clemmons DR (2005a) DOK-1 mediates SHP-2 binding to the alphaVbeta3 integrin and thereby regulates insulin-like growth factor I signaling in cultured vascular smooth muscle cells. J Biol Chem 280:3151–3158

Ling Y, Maile LA, Lieskovska J, Badley-Clarke J, Clemmons DR (2005b) Role of SHPS-1 in the regulation of insulin-like growth factor I-stimulated Shc and mitogen-activated protein kinase activation in vascular smooth muscle cells. Mol Biol Chem 16:3353–3364

Maile LA, Clemmons DR (2002) Regulation of insulin-like growth factor I receptor dephosphorylation by SHPS-1 and the tyrosine phosphatase SHP-2. J Biol Chem 277:8955–8960

Maile LA, Busby WH, Sitko K, Capps BE, Sergent T, Badley-Clarke J, Clemmons DR (2006a) Insulin-like growth factor-I signaling in smooth muscle cells is regulated by ligand binding to the 177CYDMKTTC184 sequence of the beta3-subunit of the alphaVbeta3. Mol Endocrinol 20:405–413

Maile LA, Busby WH, Sitko K, Capps BE, Sergent T, Badley-Clarke J, Ling Y, Clemmons DR (2006b) The heparin binding domain of vitronectin is the region that is required to enhance insulin-like growth factor-I signaling. Mol Endocrinol 20:881–892

Margariti A, Zeng L, Xu Q (2006) Stem cells, vascular smooth muscle cells and atherosclerosis. Histol Histopathol 21:979–985

Miyamoto S, Teramoto H, Gutkind JS, Yamada KM (1996) Integrins can collaborate with growth factors for phosphorylation of receptor tyrosine kinases and MAP kinase activation: roles of integrin aggregation and occupancy of receptors. J Cell Biol 135:1633–1642

Nichols TC, du Laney T, Zheng B, Bellinger DA, Nickols GA, Engleman W, Clemmons DR (1999) Reduction in atherosclerotic lesion size in pigs by alphaVbeta3 inhibitors is associated with inhibition of insulin-like growth factor-I mediated signaling. Circ Res 85:1040–1045

Oshima K, Ruhul Amin AR, Suzuki A, Hamaguchi M, Matsuda S (2002) SHPS-1, a multifunctional transmembrane glycoprotein. FEBS Lett 519:1–7

Ozawa K, Miyazaki M, Matsuhisa M, Takano K, Nakatani Y, Hatazaki M, Tamatani T, Yamagata K, Miyagawa J, Kitao Y, Hori O, Yamasaki Y, Ogawa S (2005) The endoplasmic reticulum chaperone improves insulin resistance in type 2 diabetes. Diabetes 54:657–663

Ozcan U, Yilmaz E, Ozcan L, Furuhashi M, Vaillancourt E, Smith RO, Gorgun CZ, Hotamisligil GS (2006) Chemical chaperones reduce ER stress and restore glucose homeostasis in a mouse model of type 2 diabetes. Science 313:1137–1140

Senger Dr, Claffey KP, Benes, JE, Perruzzi CA, Sergiou AP, Detmar M (1997) Angiogenesis promoted by vascular endothelial growth factor: regulation through alpha1beta1 and alpha2beta1 integrins. Proc Natl Acad Sci USA 94:13612–13617

Stenina OI, Krukovets I, Wang K, Zhou Z, Forudi F, Penn MS, Topol EJ, Plow EF (2003) Increased expression of thrombospondin-1 in vessel wall of diabetic zucker rat. Circulation 107:3209–3215

Zhang M, Smith EP, Kuroda H, Banach W, Chernausek SD, Fagin JA (2002) Targeted expression of a protease-resistant IGFBP-4 mutant in smooth muscle of transgenic mice results in IGFBP-4 stabilization and smooth muscle hypotrophy. J Biol Chem 277:21285–21290

Zheng B, Clemmons DR (1998) Blocking ligand occupancy of the alphaVbeta3 integrin inhibits insulin-like growth factor-I signaling in vascular smooth muscle cells. Proc Natl Acad Sci USA 95:11217–11222
Zhu B, Zhao G, Witte DP, Hui DY, Fagin JA (2001) Targeted overexpression of IGF-I in smooth muscle cells of transgenic mice enhances neointimal formation through increased proliferation and cell migration after intraarterial injury. Endocrinology 142:3598–3606

Deciphering the Conundrum of Estrogen-driven Breast Cancer: Aurora Kinase Deregulation

Jonathan J. Li[1] and *Sara Antonia Li*[1]

Abstract

Of all cancers occurring in women in developed countries, breast cancer alone accounts for about 32%. This is obviously an important world-wide public health concern. More than 90% of all human breast cancer cases are sporadic or non-familial, with an equally high percentage of these cases being ductal carcinomas. Estrogens have a profound role in the causation and progression of human sporadic breast cancer. In premenopausal woman, all of the well-established risk factors clearly implicate estrogens in the etiology of sporadic ductal breast cancer. This occurs within a narrow range of serum and breast tissue levels of 17β-estradiol (E_2) concentrations, all in the low picogram range. Although estrogen receptor α (ERα) is commonly found and is a highly important characteristic of human sporadic breast cancer, it is not this cancer's most defining feature. Rather, the hallmarks of primary invasive ductal breast neoplasms as well as ductal carcinoma in situ (DCIS), a pre-invasive premalignant lesion, are chromosomal instability and aneuploidy. These molecular characteristics have been reported in 55–78% of the DCISs and in 85–92% of invasive disease. So, how does one go from estrogen via ERα to aneuploidy in breast cells? While all the pieces of this puzzle are not in place, some of the molecular alterations that occur during E_2-induced breast oncogenesis may be predicted whereas others are not so apparent. On the basis of our studies, we have developed a new paradigm in a murine breast tumor model in female ACI rats induced by near physiological levels of estrogen. This mammary tumor animal model exquisitely resembles human sporadic ductal breast cancer, both in its histopathologic and its molecular progression. Estrogen, interacting with the ERα, transactivates the over-expression of c-*myc* and its eventual amplification as a result of gains in chromosome 8 in the human and chromosome 7 in the rat (Andrieu et al. 2000). Subsequent deregulation of specific entities in the cell cycle, including cyclin D1, E1, their respective binding partners cdk4 and cdk2, and MDM2, is notable in both human and murine pre-malignant breast lesions and frank tumors (Clemons and Goss 2001). Unexpectedly, Aurora A and B, members of a mitotic kinase family, were found to be persistently over-expressed in E_2-induced ACI rat mammary tumors, a finding that has also been reported in sporadic human ductal breast cancers (94%) (Andrieu

---

[1] Hormonal Oncogenesis Laboratory, Department of Pharmacology, Toxicology and Therapeutics, The University of Kansas Medical Center, 3901 Rainbow Boulevard, Kansas City, KS, 66160, USA
*e-mail:*jli1@kumc.edu

Melmed et al.
Hormonal Control of Cell Cycle
© Springer-Verlag Berlin Heidelberg 2008

et al. 2000; Feigelson and Henderson 1996). Both rodent and human breast neoplasias also exhibited centrosome amplification at high frequency (>80%) in pre-malignant lesions and in primary tumors. This amplification presumably occurs as a consequence of centrosome duplication and possibly separation errors elicited by Aurora A and B kinase deregulation of mitotic protein substrate phosphorylations. Thus, estrogen-driven deregulations of the cell cycle and the mitotic machinery are crucial molecular events leading to aneuploidy and ultimately to breast cancer in this E_2-induced murine mammary tumor model and in women.

Introduction

Breast cancer (BC) is the most commonly diagnosed malignancy in developed countries, with 1.2 million new cases yearly worldwide (Parkin et al. 2005). The incidence of BC essentially equals the combined incidences of lung, colorectal, ovarian, and endometrial cancers in women (Jemal et al. 2005); therefore, it is clearly an immense public health issue. The vast majority of all BCs ($\sim$90%) are sporadic or non-familial (Andrieu et al. 2000) and an equally high percentage of all cases are ductal BCs, the remainder being lobular BCs (10–15%).

More than 300 years have elapsed since Ramazzi, an Italian physician, raised a conundrum when he reported that nuns had a higher incidence of BC compared to women in the general population. This observation has been extended in the past 75 years in both animal and human studies, with the identification of ovarian hormones, particularly estrogens, as the major etiologic agent in elevating BC risk (Clemons and Goss 2001). This finding has led to a compilation of well-established BC risk factors that are all related to higher and sustained estrogen serum levels found in normal cycling women (Fig. 1).

These BC risk factors include early first menarche, late age of menopause, nulliparity, late age at full-term pregnancy, and absence of lactation (Clemons and Goss 2001; Feigelson and Henderson 1996; Adami et al. 1995), all related to pre-menopausal women. In post-menopausal women, obesity and use of combined hormone replacement therapy but not estrogen replacement therapy are risk factors. Despite this knowledge, a precise cellular and molecular understanding of how estrogens affect BC risk has so far remained elusive. In human BC causation, there are certain inherent considerations that must be taken into account if one is to understand the mecha-

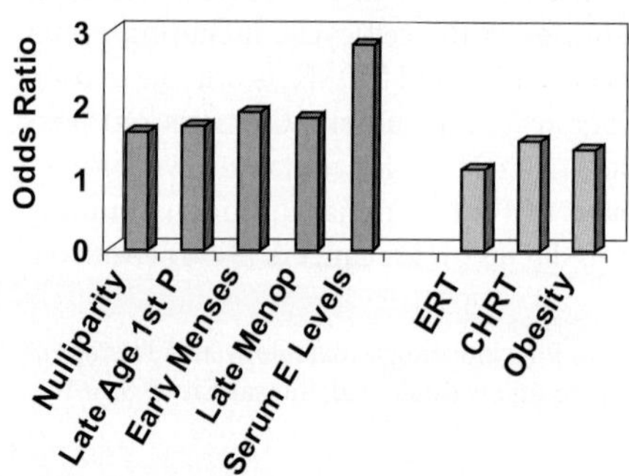

Fig. 1. Established risk factors for sporadic breast cancer. P, pregnancy; ERT, estrogen replacement therapy; CHRT, combined hormone replacement therapy

nism by which estrogens elicit their oncogenic changes in breast tissue. One is the relatively low concentrations of 17β-estradiol (E_2) in the serum and breast tissue in normal cycling women, all in the low pg/ml and pg/mg of protein, respectively. The second is the most defining characteristic of ductal BC, that is, the tumors are highly aneuploid.

Estrogen Concentrations and Breast Cancer

The relationship between endogenous estrogens and BC risk has largely been studied in postmenopausal women, with only a few small studies performed in pre-menopausal women (Toniolo 1997). In a recently published, nested case-control study within the Nurses' Health Study II, plasma E_2 levels were determined in a large group of pre-menopausal women during the early follicular and mid-luteal phases of their menstrual cycle (Eliassen et al. 2006). Women in the highest quartiles of the follicular phase had total and free E_2 levels of 66–100 and 0.8–1.2 pg/ml, respectively, and showed significantly elevated BC risk, RR = 2.1 and 2.4, respectively. This association was stronger for invasive BC tumors and for estrogen and progesterone receptor positive (ER^+/PR^+) BCs. However, luteal E_2 levels were not associated with an increased BC risk. Higher testosterone and androstenedione levels in both menstrual cycle phases exhibited a modest association of increased BC risk, which was stronger for ER^+/PR^+ tumors. On the other hand, the levels of estrone (E_1), estrone sulfate, progesterone, and sex hormone binding globulin were not associated with an increased BC risk in this study.

The circulating level of E_2 over a woman's lifetime, excluding pregnancy, is one of the most important parameters in relation to BC risk. Total serum E_2 levels are related to the phase of the menstrual cycle, the output of E_2 from the ovary and, to some extent, the synthesis/conversion of E_2 in peripheral tissues and in the breast itself. Over a woman's lifetime, the level of E_2 serum in non-pregnant pre-menopausal women ranges from 30–360 pg/ml (Becker 1995), whereas in post-menopausal women it varies between 5–35 pg/ml (Becker 1995; Colditz 1998; Probst-Hensch et al. 2000). The concentration of E_2 and E_1 in normal human breast tissue is 5.7 and 3.9 pg/mg protein, respectively (Chetrite et al. 2000; Vermeulen et al. 1986), whereas in BC tissue these levels are moderately higher, 8.9 and 6.8 pg/mg protein, respectively. There is considerable individual variability in E_2 concentrations in human normal breast and breast tumor tissue. It has been reported that post-menopausal breast and breast tumor tissue E_2 and E_1 concentrations fluctuate between 10–20- and 2–10-fold higher, respectively, than corresponding serum estrogen levels (Clarke et al. 2001). However, assuming that these E_2 and E_1 values are valid, it would not be expected that breast tissue estrogen levels of most post-menopausal women would commonly attain concentrations of 0.2–0.3 ng/mg protein.

In female ACI rats, the serum E_2 concentration sufficient to induce a high incidence of breast tumors varies between 60 and 120 pg/ml, a range approaching the high end of the physiological serum E_2 levels (10–45 pg/ml) in normal cycling rats (Naftolin et al. 1972).

Aneuploidy

Human solid neoplasms and many hematological malignancies are commonly aneuploid, that is, a change in whole chromosome gains/losses. However, the aneuploid frequencies among solid tumors, from different organ sites, are not always similar. For a few solid tumors, namely breast, kidney, and bladder, aneuploidy is perhaps their most defining characteristic (Li and Li 2006), with mean percent aneuploid frequencies of 78, 75, and 80%, respectively. Significantly, high aneuploid frequencies have also been detected in early stages (carcinomas in situ) in the development of at least two (breast and bladder) of these three tumors. The detection of aneuploidy in pre-malignant stages strongly implicates the involvement of this alteration in the oncogenic process itself and not only in the progression of the established tumors.

For nearly half a century, the predominant animal models in BC research have utilized various synthetic chemical carcinogens (i.e., 7,12-dimethyl benz[α]anthracene (DMBA), nitrosomethylurea (NMU)), none of which are found in the environment. We and others have shown that the breast tumors induced in female BUF/N and SD rats by DMBA, NMU and 6-nitrochrysene (6-NC), an environmental carcinogen, are largely diploid (85–90%; Li et al. 2002; Aldaz et al. 1992; Haag et al. 1996). In contrast, breast tumors induced in female ACI or Noble rats by E_2 alone or in combination with testosterone, respectively, were highly aneuploid (89–91%; Li et al. 2002). Moreover, pre-malignant lesions, ductal carcinoma in situ (DCIS), from these hormone-induced models were also highly aneuploid (>85%). It is evident from these findings that the E_2-induced breast tumors in ACI rats more closely resemble human BC in this pertinent feature than do mammary tumors induced by synthetic chemical carcinogens.

Histopathology

A relevant characteristic of E_2-induced breast tumors in female ACI rats is the sequence of morphologic changes elicited by relatively low, albeit constant, E_2 serum levels. Hyperplasia of breast epithelial cells is an expected occurrence when the mammary gland is exposed to constant or periodic estrogen and progesterone; both hormones are mitogenic in this tissue. However, initial studies in our laboratory indicate that the hyperplasia elicited by estrogen and endogenous progesterone in intact female ACI rats does not appear to give rise to the ductal mammary tumors that eventually develop. Rather, the earliest focal dysplasias elicited by chronic low dose estrogen treatment are clusters of cells that contain large pale-staining nuclei (Fig. 2B), distinct from the normal proliferating epithelial cells driven by both female hormones. With continued E_2 treatment, the pre-malignant stages remarkably resemble the histopathology commonly seen in the development of human invasive ductal BC. In the ACI rat model, we have designated three stages of dysplasia (Fig. 2B), with stage 3 resembling atypical ductal hyperplasia (ADH). At 4.3 months of E_2 treatment, DCISs begin to appear (Fig. 2C–E). The DCIS most commonly seen in human sporadic breast premalignancies is the cribiform type, also seen in E_2-treated ACI rats (Fig. 2C); this type is 8 to 10 times more common than the comedo type (Fig. 2D). The other DCIS

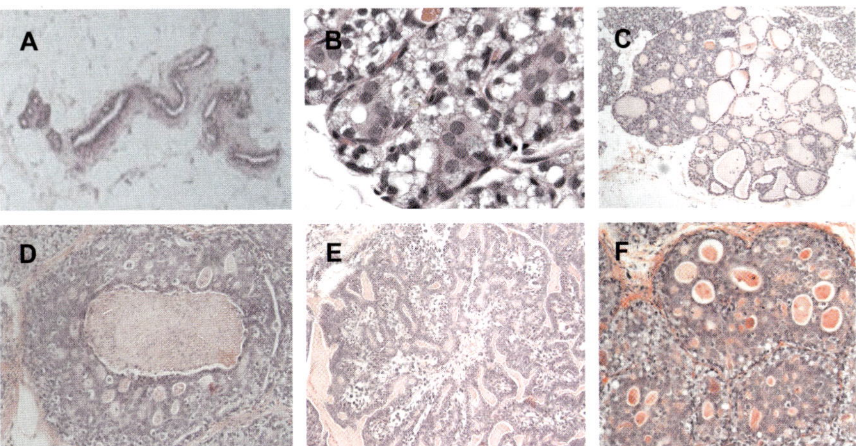

Fig. 2. Female ACI rats. **A** Control untreated mammary gland. **B** Incipient lesion, 3.0 mo E_2. **C** Cribiform DCIS, 4.5 mo E_2. **D** Comedo DCIS, 5.0 mo E_2. **E** Papillary DCIS, 5.0 mo E_2. **F** Invasive ductal BC, 6.0 mo E_2. Magnification $\times 40$

types, solid and papillary (Fig. 2E), are found less frequently, similar to human breast pre-malignant frequencies.

Steroid Receptors

Of the sporadic BC cases, $55-73\%$ are ERα positive. Although the native 66-kDa is the major form, its splice variant 46-kDa has been detected, and both forms are considered ligand-activated transcription factors that regulate transcription of estrogen-responsive genes in the cell nucleus (Weihua et al. 2003; Kong et al. 2003). Recently, a human 36-kDa ERα variant has been cloned that lacks both transcriptional activation domains (AF-1 and AF-2) but retains the DNA-binding domain of the native ERα (Wang et al. 2005).

In female ACI rat breast tissue, E_2 induces the expression of a number of ERα forms, including the native full-length 66-kDa and the variants 47-, 54-, and 56-kDa (Li et al. 2002). The expression of the native 66-kDa ERα was predominant in E_2-induced breast tissue and in normal E_2-stimulated uterus but was not the primary form in ACI rat primary breast tumors (Li et al. 2002). The 47-kDa ERα variant was detected only in E_2-induced breast tissues but not in tumors. The 54-kDa ERα variant was identified in primary ACI rat breast tumors and in the E_2-stimulated uterus, whereas the 56-kDa ERα variant was detected in age-matched untreated controls, Tamoxifen (TAM)- and E_2-treated proliferating breast tissues, and primary breast tumors (Li et al. 2002). Progesterone receptors A, B, and C were also detected in E_2-treated mammary glands, primary breast tumors, and uterine tissue. However, only PR-C was found in control untreated and TAM-treated mammary glands. High frequencies of ERα- and PR-positive cells were detected in all pre-malignant stages and in E_2-induced primary breast neoplasms (Li et al. 2002).

However, it has not been resolved which ERα forms reside in early pre-malignant stages.

Cell Cycle

A common characteristic of sporadic human ductal BC is deregulation of the cell cycle, particularly cyclin D1 and E1, as well as their respective kinase binding partners, cdk 4 and cdk 2 (Xiong et al. 1991; Sutherland and Musgrove 2004; Steeg and Zhou 1998). This deregulation is expected to confer a growth advantage to pre-malignant and malignant BC stages. The over-expression of these cyclins is mediated in large part by the upregulation of *c-myc* and its protein product as a result of E-ERα action, a common feature in human ductal BCs (Dubik et al. 1987). Since the ERα response element can also directly interact with the cyclin D1 promoter, it can effectively bypass c-MYC (Petrizzi et al. 2001).

Our laboratory has found a 24.0-fold elevation in cyclin D1 expression in E_2-induced ACI rat breast tumors and a concomitant 3.0-fold rise in cdk4 (Fig. 3A) compared to untreated age-matched animals. Employing the Rb kinase assay, cyclin D1 activity in the breast tumors was significantly elevated, as was its specific binding to cdk4

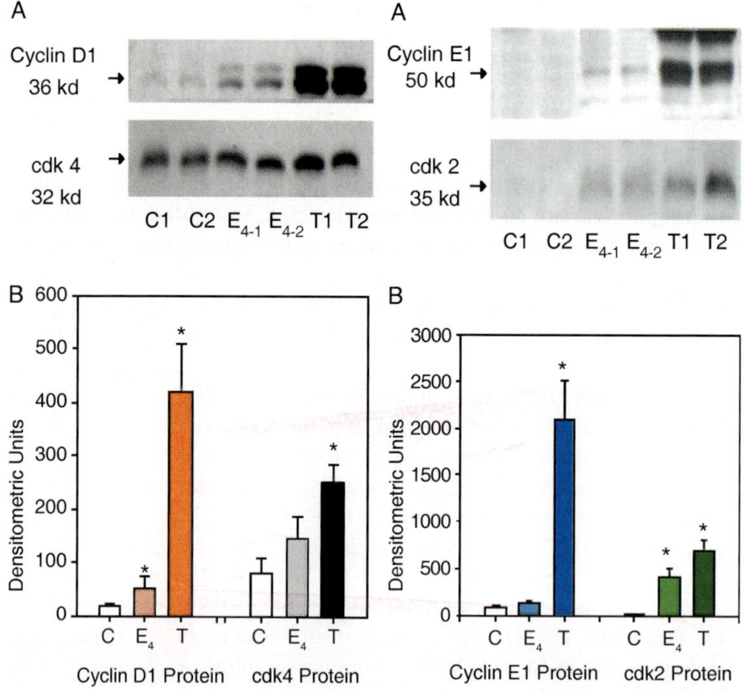

Fig. 3. In vivo complex formation between cyclin D1 and E1 with cdk4 and ckd2, respectively. *t-Test, significant if $p < 0.05$

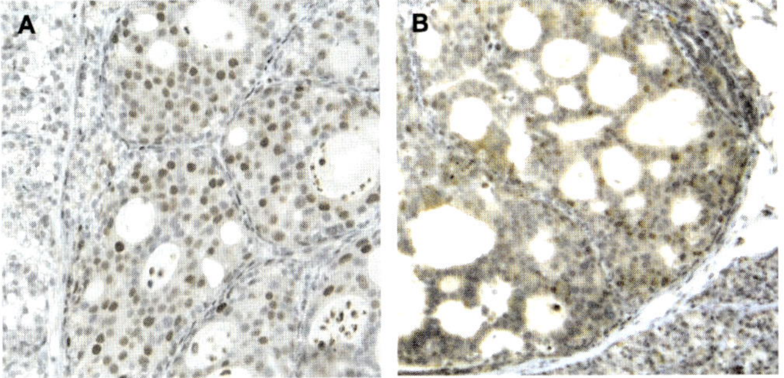

Fig. 4. Immunohistochemical localization of cyclin D (**A**) and cyclin E1 (**B**) in ACI rat E_2-induced mammary gland tumors

(Weroha et al. 2006). Furthermore, PCR analysis showed a 2.9-fold rise in cyclin D1 mRNA expression, albeit no amplification of the cyclin D1 gene was detected (Weroha et al. 2006). Similarly, cyclin E1 expression increased 24.0-fold in primary ACI rat mammary tumors, together with a 45.0-fold rise in cdk2 (Fig. 3B). Using histone H1 as a substrate, cyclin E1 activity significantly increased and its specific binding to cdk 2 was established (Weroha et al., in preparation). In the same study, cyclin E1 mRNA expression showed a 2.1-fold elevation, and 20% of the ACI rat breast tumors examined exhibited gene amplification of this cyclin.

Dysplastic foci, DCIS (Fig. 4) and primary ACI rat mammary tumors showed a preferential and significant increase in cyclin D1 and E1 expression when compared to normal hyperplasia-elicited E_2 treatment (Weroha et al. 2006; in preparation). It is evident from these data that there is a growth advantage in E_2 pre-malignant lesions. Additionally, an equally if not more important role for cyclin E1.cdk2 can be ascribed to its involvement in eliciting chromosomal instability and aneuploidy in vitro (Spruck et al. 1999). However, the precise mechanisms and interaction of the individual entities involved need elucidation.

Aurora Kinases

Aneuploidy has been widely employed as a diagnostic marker in human breast DCISs and in primary BCs for many decades. However, it has not been realized until recently that it provides an important clue to the causation of sporadic human BC and the role of estrogen in its etiology (Li et al. 2004). As Boveri predicted more than a century ago (Boveri 1914), deregulation of the centrosome cycle leads to supernumerary centrosomes and multipolar spindles during mitosis, an essential mechanism whereby oncogenesis occurs. How then would this process take place? While the main pieces of this conundrum are beginning to fall into place, the mechanistic details remain lacking.

The Aurora family of kinases in mammalian cells consists of three members: Aur-A, -B, and -C. The Auroras are part of a mitotic kinase superfamily that includes polo-like, Nek (NIMA related), and cdk1 (Kramer et al. 2004), the latter being involved in the mitotic spindle checkpoint (Li and Li 2006). Focusing our discussion on the Aur family, it has been established that these kinases are critically involved in centrosome duplication, maturation and separation, spindle assembly and stability, chromosome condensation and segregation, and cytokinesis. Over-expressed Aur-A has been shown to elicit neoplastic transformation in mammalian cells, both in vitro and in vivo (Brinkley and Goepfert 1998; Zhou et al. 1998), indicative of its oncogenic potential. Aurora kinases perform and coordinate, overlapping and distinctive functions during mitosis (Andrews et al. 2003). These serine/threonine kinases operate in concert to regulate the tightly control processes of normal cell division. Aur-A is localized to the centrosome. In late G1/early S phase, Aur-A is localized to the pericentriolar material of the centrosomes, and as the cell cycle proceeds, its level of expression rises (Marumoto et al. 2005). Continuing its association with centrosomes at the mitotic poles in prophase, it then resides in adjacent spindle microtubules in metaphase. Aur-B and Aur-C are passenger proteins localized between sister centrosomes on the inner centromere from late G2 through metaphase. Aur-B is concentrated in the spindle midzone and in the cell cortex at the site of cleavage-furrow ingression. Aur-B functions in condensation, segregation, and cytokinesis by regulating microtubule kinetochore associations. Aur-C is a passenger protein whose functions are not well known (Sasai et al. 2004). Collectively, the Aurora kinase family phosphorylates about 20 centrosomal and mitotic protein substrates required for proper cell division. A major challenge is to determine the precise role of Aurora substrate alterations when sustained Aurora kinase over-expressions occur and their involvement in the generation of aneupoidy and oncogenesis in BC, as well as in cancers at other sites.

Over-expressed Aur-A protein has been detected in 94% of the human BCs examined (Tanaka et al. 1999). In addition, it has been reported that 62% of human BC samples exhibited either high or intermediate Aur-A mRNA expression levels (Miyoshi et al. 2001). Recently, markedly higher levels of Aur-A expression were observed in hu-

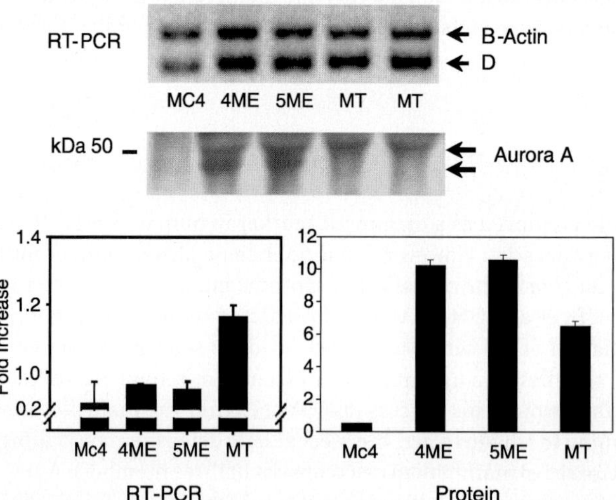

Fig. 5. RT-PCR and protein expression of Aur-A during E_2-induced oncogenesis in the ACI rat mammary gland

man breast DCISs (Hoque et al. 2003) compared to primary breast tumors, indicating its early involvement in human BC oncogenesis.

In 4.0-month E_2 treated and primary mammary tumors, Aur-A mRNA was elevated 1.4- and 1.5-fold, respectively, compared to age-matched untreated controls (Li et al. 2004), while its protein expression rose 7.2−7.4-fold. Recently, we have detected a > 10.0-fold rise in Aur-A protein expression after 4.3−4.5 months of E_2 treatment in mammary glands of ACI rats (Fig. 5). This finding is noteworthy since this increase occurs during the E_2-treatment period where the occurrence of DCISs is maximal. The over-expression of Aur-A is largely confined to cells within focal dysplasias, DCISs, and primary breast tumors elicited by E_2 in this murine model.

Amplified Centrosomes

For proper cell division, the centrosome functions as the microtubule organizing center (MTOC) for the nucleation of microtubule arrays. During mitosis, the centrosome has an essential role in the equal segregation of chromosomes by the establishment of the bipolar spindle (Salisbury 2001). Centrosome defects have been implicated as a primary cause of chromosomal instability and aneuploidy in cancer causation (Boveri 1914; Lingle et al. 1998; Brinkley 2001). Centrosome amplification is characterized by an increase in centrosome size and volume as a consequence of accumulated γ-tubulin, centrin, and pericentrin; comprising the pericentriolar matrix. There is also an increase in centrosome number, elevated microtubule nucleation capacity, and increased levels of phosphorylated centrosomal and other mitotic-associated proteins (Lingle et al. 1998, 2001).

A high frequency ($\sim$ 80%) of centrosome amplification has been reported in human ductal BC (Lingle et al. 1998, 2002), with individual human breast tumor cells possessing three to eight centrosomes. Moreover, in addition to centrosome number, both centrosome size and volume exhibited a positive linear correlation with chromosomal instability and aneuploidy. Amplified centrosomes have also been detected in human

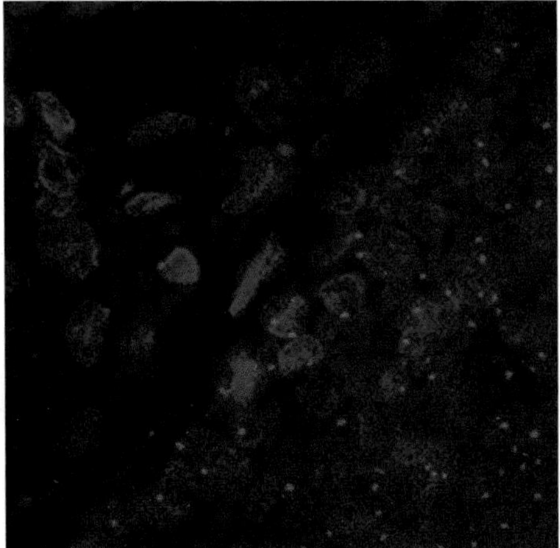

Fig. 6. Centrosome amplification in an E_2-induced female ACI rat mammary tumor (lower right). *Blue* = nuclear material, *red* = γ-tubulin

breast DCISs. These seminal findings have several important implications. First, the detection of amplified centrosomes in pre-malignant human BC stages suggests that it is likely an early crucial event in the oncogenic process. Second, the consequent development of chromosomal instability would foster heterogeneity in tumor cell populations and tumor phenotypic aggressivity.

Remarkably similar is the finding that amplified centrosomes are ubiquitous in E_2-induced primary ACI ductal mammary tumors (Fig. 6) and in DCISs (Li et al. 2004). Of particular interest is the finding that ACI rat focal dysplasias, present after a 3.0-month E_2 treatment, exhibit amplified centrosomes. Consistent with this finding is a 2.4-fold rise in γ-tubulin and an 11.2-fold increase in centrin in ACI rat breast tumors (Li et al. 2004). These data now link deregulated cell cycle entities, sustained Aur-A/-B over-expression, and centrosome amplification with estrogen in mammary oncogenesis. It is striking that these same alterations occur in high frequency in human ductal pre-malignant and invasive BC stages.

Chromosomal Instability

There are no sufficiently large studies accurately identifying the frequency of chromosomal gains/losses in aneuploid human sporadic ductal BCs and the consequent amplification of specific individual genes. Nevertheless, Erb-B2 (Her-2/neu), hst/int-2, cyclins D1 and E1, and cdk4 are some of the genes reported to exhibit low but relatively consistent amplification frequencies in invasive ductal BCs (Adnane et al. 1989). In different studies of sporadic BC, c-*myc* amplification has been shown, varying from 1 to 94% frequency (Deming et al. 2000). The wide range of frequencies reported for amplified c-*myc* in human BC is due in part to the inconsistencies of the assay method used by different groups. Few studies have examined the simultaneous occurrence of multiple gene amplifications in a single cohort of aneuploid sporadic ductal BCs (Adnane et al. 1989). In another study, c-*myc* was the most common gene amplified in human ductal BCs, and log linear analysis indicated that it was the first gene amplified among four examined (Janocko et al. 1995). We performed extensive G-banded karyotypic and comparative genomic hybridization (CGH) analyses in E_2-induced ACI rat aneuploid breast tumors (Fig. 7; Li et al. 2004). Non-random or consistent chromosome gains, primarily trisomies and to a lesser extent tetrasomies, were seen in

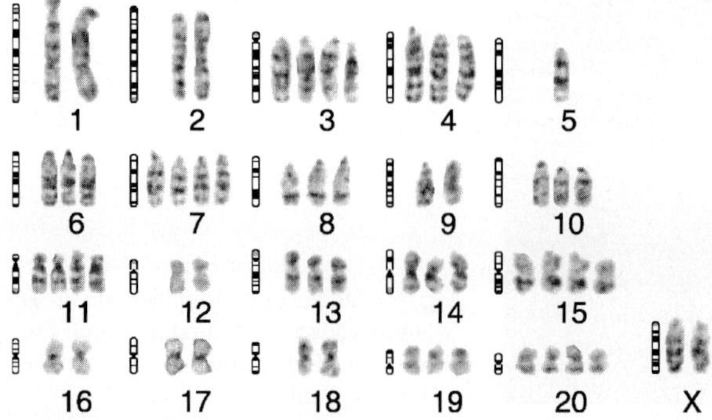

Fig. 7. Representative Giemsa-banded karyotype by an ACI rat 5.5-mo E_2-induced mammary tumor. Note trisomies in chromosomes 4, 6, 8, 10, 13, and 19 and tetrasomies in chromosomes 3, 7, 11, 15, and 20

chromosomes 7, 11, 12, 13, 19, and 20 and a single consistent loss was seen in chromosome 12 (Fig. 7). Using a fluorescent c-*myc* specific probe, the c-*myc* gene was localized on chromosome 7q33 in the ACI rat. Amplification of c-*myc* (3.4−6.9 copy number) occurred in E_2-induced primary rat breast tumors with a frequency of 66%. It is likely that some of the genes residing on other non-randomly gained chromosomes will prove important in contributing to the breast oncogenesis in this model. The CGH analysis of E_2-induced primary ACI rat breast tumors largely coincided with the numerical chromosome alterations. Taken together, these data are consistent with the belief that mammary tumors are clonally derived by an estrogen-driven process.

Summary and Future Directions

A new paradigm is presented that provides a sequence of molecular changes, initiated and driven by estrogen in breast oncogenesis in the female ACI rat (Fig. 8), that is striking in its resemblance to similar pertinent alterations found in human breast pre-malignancy and primary sporadic invasive ductal BCs. In susceptible mammary cells, E_2 interacts with its receptor, ERα, to elicit c-*myc*/MYC over-expression. As a consequence of MYC overexpression, certain cell cycle-related gene/protein entities are over-expressed, i.e., cyclins D1, E1 and their respective cdks, and MDM2, downstream. E-ERα may directly activate cyclin D1. Additionally, due in part to the over-expressed cell cycle protein entities, persistent Aurora kinase (A/B) expression occurs. It is possible that E-ERα also interacts directly with the Aurora kinase gene to enhance Aur-A/-B protein. Persistently, over-expressed Aur-A is considered a mutational event resulting in hyper-phosphorylation of centrosomal mitotic protein substrates, which would likely result in severe centrosome cycle disturbances, primarily in centrosome duplication, but also possibly in separation. Centrosome amplification will lead to the generation

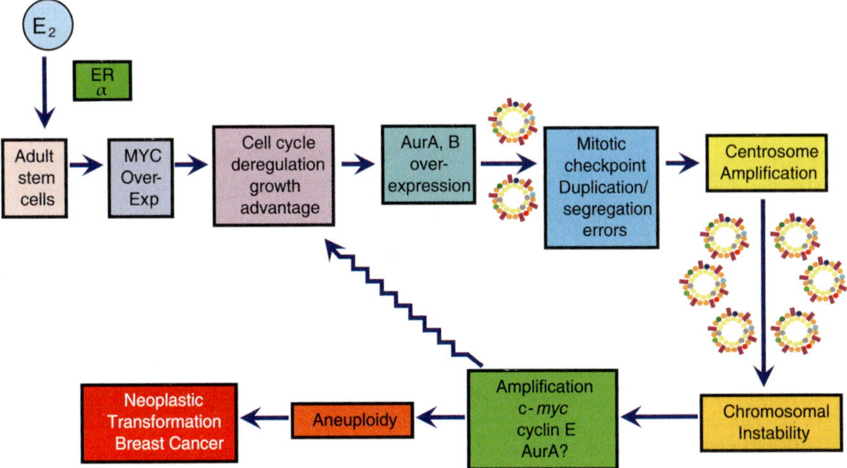

Fig. 8. A novel paradigm for estrogen-driven breast oncogenesis

of multipolar spindles and the missegregation of chromosomes. If this sequence is correct, it should provide new insight into BC causation, by defining an etiologic role for estrogens at essentially physiological serum concentrations. However, there are numerous areas at the mechanistic level that need to be addressed: first, an understanding of the association between E-ERα and Aurora kinases; second, elucidation of the relationship between deregulated cell cycle entities and Aurora kinase over-expression; third, clarification of the link between persistent Aurora kinase over-expression and the generation of amplified centrosomes; and fourth, the generation of chromosomal instability and aneuploidy in relation to neoplastic transformation. In conclusion, based on this sequence of estrogen-induced breast oncogenesis, new molecular targets are envisioned that would provide novel strategies not only for BC treatment but also for its prevention.

Acknowledgements. We thank Dr. Wilma Lingle, Mayo Clinic, for providing the centrosome (Fig. 6) and Dr. John S. Weroha for the Aurora kinase (Fig. 5) for this review.

References

Adami HO, Persson I, Ekbom A, Wolk A, Ponten J, Trichopoulos D (1995) The aetiology and pathogenesis of human breast cancer. Mutat Res 333:29–35

Adnane J, Gaudray P, Simon MP, Simony-Lafontaine J, Jeanteur P, Theillet C (1989) Proto-oncogene amplification and human breast tumor phenotype. Oncogene 4:1389–95

Aldaz CM, Gollahon LS, Chen A (1992) Chromosome alterations in rat mammary tumor progression. Prog Clin Biol Res 376:137–153

Andrews PD, Knatko E, Moore WJ, Swedlow JR (2003) Mitotic mechanics: the auroras come into view. Curr Opin Cell Biol 15:672–683

Becker K (1995) Principles and practice of endocrinology and metabolism. Ed 2. Lippincott Co, Philadelphia, PA

Boveri T (1914) The origin of malignant tumours. Williams and Wikins, Baltimore

Brinkley BR (2001) Managing the centrosome numbers game: from chaos to stability in cancer cell division. Trends Cell Biol 11:18–21

Brinkley BR, Goepfert TM (1998) Supernumerary centrosomes and cancer: Boveri's hypothesis resurrected. Cell Motil Cytoskeleton 41:281–288

Chetrite GS, Cortes-Prieto J, Philippe JC, Wright F, Pasqualini JR (2000) Comparison of estrogen concentrations, estrone sulfatase and aromatase activities in normal, and in cancerous, human breast tissues. J Steroid Biochem Mol Biol 72:23–27

Clarke R, Leonessa F, Welch JN, Skaar TC (2001) Cellular and molecular pharmacology of antiestrogen action and resistance. Pharmacol Rev 53:25–71

Clemons M, Goss P (2001) Estrogen and the risk of breast cancer. N Engl J Med 344:276–285

Colditz GA (1998) Relationship between estrogen levels, use of hormone replacement therapy, and breast cancer. J Natl Cancer Inst 90:814–823

Deming SL, Nass SJ, Dickson RB, Trock BJ (2000) C-myc amplification in breast cancer: a meta-analysis of its occurrence and prognostic relevance. Br J Cancer 83:1688–1695

Dubik D, Dembinski TC, Shiu RP (1987) Stimulation of c-myc oncogene expression associated with estrogen-induced proliferation of human breast cancer cells. Cancer Res 47:6517–6521

Eliassen AH, Missmer SA, Tworoger SS, Hankinson SE (2006) Endogenous steroid hormone concentrations and risk of breast cancer: does the association vary by a woman's predicted breast cancer risk? J Clin Oncol 24:1823–1830

Feigelson H, Henderson B (1996) Estrogens and breast cancer. Carcinogenesis 17:2279–2284

Haag JD, Hsu LC, Newton MA, Gould MN (1996) Allelic imbalance in mammary carcinomas induced by either 7,12-dimethylbenz[a]anthracene or ionizing radiation in rats carrying genes conferring differential susceptibilities to mammary carcinogenesis. Mol Carcinog 17:134–143

Hoque A, Carter J, Xia W, Hung MC, Sahin AA, Sen S, Lippman SM (2003) Loss of aurora A/STK15/BTAK overexpression correlates with transition of in situ to invasive ductal carcinoma of the breast. Cancer Epidemiol Biomarkers Prev 12:1518–1522

Janocko LE, Lucke JF, Groft DW, Brown KA, Smith CA, Pollice AA, Singh SG, Yakulis R, Hartsock RJ, Shackney SE (1995) Assessing sequential oncogene amplification in human breast cancer. Cytometry 21:18–22

Jemal A, Murray T, Ward E, Samuels A, Tiwari RC, Ghafoor A, Feuer EJ, Thun MJ (2005) Cancer statistics, 2005. CA Cancer J Clin 55:10–30

Kong EH, Pike AC, Hubbard RE (2003) Structure and mechanism of the oestrogen receptor. Biochem Soc Trans 31:56–59

Kramer A, Mailand N, Lukas C, Syljuasen RG, Wilkinson CJ, Nigg EA, Bartek J, Lukas J (2004) Centrosome-associated Chk1 prevents premature activation of cyclin-B-Cdk1 kinase. Nature Cell Biol 6:884–891

Li JJ, Li SA (2006) Mitotic kinases: the key to duplication, segregation, and cytokinesis errors, chromosomal instability, and oncogenesis. Pharmacol Ther 111:974–984

Li JJ, Papa D, Davis MF, Weroha SJ, Aldaz CM, El-Bayoumy K, Ballenger J, Tawfik O, Li SA (2002) Ploidy differences between hormone- and chemical carcinogen-induced rat mammary neoplasms: comparison to invasive human ductal breast cancer. Mol Carcinog 33:56–65

Li JJ, Weroha SJ, Lingle WL, Papa D, Salisbury JL, Li SA (2004) Estrogen mediates Aurora-A overexpression, centrosome amplification, chromosomal instability, and breast cancer in female ACI rats. Proc Natl Acad Sci USA 101:18123–18128

Li SA, Weroha SJ, Tawfik O, Li JJ (2002) Prevention of solely estrogen-induced mammary tumors in female aci rats by tamoxifen: evidence for estrogen receptor mediation. J Endocrinol 175:297–305

Lingle WL, Salisbury JL (2001) Methods for the analysis of centrosome reproduction in cancer cells. Methods Cell Biol 67:325–336

Lingle WL, Lutz WH, Ingle JN, Maihle NJ, Salisbury JL (1998) Centrosome hypertrophy in human breast tumors: implications for genomic stability and cell polarity. Proc Natl Acad Sci USA 95:2950–2955

Lingle WL, Barrett SL, Negron VC, D'Assoro AB, Boeneman K, Liu W, Whitehead CM, Reynolds C, Salisbury JL (2002) Centrosome amplification drives chromosomal instability in breast tumor development. Proc Natl Acad Sci USA 99:1978–1983

Marumoto T, Zhang D, Saya H 2005 Aurora-A – a guardian of poles. Nat Rev Cancer 5:42–50

Miyoshi Y, Iwao K, Egawa C, Noguchi S (2001) Association of centrosomal kinase STK15/BTAK mRNA expression with chromosomal instability in human breast cancers. Int J Cancer 92:370–373

Naftolin F, Brown-Grant K, Corker CS (1972) Plasma and pituitary luteinizing hormone and peripheral plasma oestradiol concentrations in the normal oestrous cycle of the rat and after experimental manipulation of the cycle. J Endocrinol 53:17–30

Parkin DM, Bray F, Ferlay J, Pisani P (2005) Global cancer statistics, 2002. CA Cancer J Clin 55:74–108 (5) Andrieu N, Prevost T, Rohan TE, Luporsi E, Le MG, Gerber M, Zaridze DG, Lifanova Y, Renaud R, Lee HP, Duffy SW (2000) Variation in the interaction between familial and reproductive factors on the risk of breast cancer according to age, menopausal status, and degree of familiality. Int J Epidemiol 29:214–223

Petrizzi V, Cicatiello L, Altucci L, Addeo R, Borgo R, Cancemi M, Ancora M, Leyva J, Bresciani F (2001) Transcriptional control of cell cycle progression by estrogenic hormones: Regulation of human cyclin D1 gene promoter activity by estrogen receptor alpha. Springer-Verlag New York, Inc, New York, NY

Probst-Hensch NM, Pike MC, McKean-Cowdin R, Stanczyk FZ, Kolonel LN, Henderson BE (2000) Ethnic differences in post-menopausal plasma oestrogen levels: high oestrone levels in Japanese-American women despite low weight. Br J Cancer 82:1867–1870

Sasai K, Katayama H, Stenoien DL, Fujii S, Honda R, Kimura M, Okano Y, Tatsuka M, Suzuki F, Nigg EA, Earnshaw WC, Brinkley WR, Sen S (2004) Aurora-C kinase is a novel chromosomal passenger protein that can complement Aurora-B kinase function in mitotic cells. Cell Motil Cytoskeleton 59:249–263

Salisbury JL (2001) The contribution of epigenetic changes to abnormal centrosomes and genomic instability in breast cancer. J Mammary Gland Biol Neoplasia 6:203–212

Spruck CH, Won KA, Reed SI (1999) Deregulated cyclin E induces chromosome instability. Nature 401:297–300

Steeg PS, Zhou Q (1998) Cyclins and breast cancer. Breast Cancer Res Treat 52:17–28

Sutherland RL, Musgrove EA (2004) Cyclins and breast cancer. J Mammary Gland Biol Neoplasia 9:95–104

Tanaka T, Kimura M, Matsunaga K, Fukada D, Mori H, Okano Y (1999) Centrosomal kinase AIK1 is overexpressed in invasive ductal carcinoma of the breast. Cancer Res 59:2041–2044

Toniolo PG (1997) Endogenous estrogens and breast cancer risk: the case for prospective cohort studies. Environ Health Perspect 105 Suppl 3:587–592

Vermeulen A, Deslypere JP, Paridaens R, Leclercq G, Roy F, Heuson JC (1986) Aromatase, 17 beta-hydroxysteroid dehydrogenase and intratissular sex hormone concentrations in cancerous and normal glandular breast tissue in postmenopausal women. Eur J Cancer Clin Oncol 22:515–525

Wang Z, Zhang X, Shen P, Loggie BW, Chang Y, Deuel TF (2005) Identification, cloning, and expression of human estrogen receptor-alpha36, a novel variant of human estrogen receptor-alpha66. Biochem Biophys Res Commun 336:1023–1027

Weihua Z, Andersson S, Cheng G, Simpson ER, Warner M, Gustafsson JA (2003) Update on estrogen signaling. FEBS Lett 546:17–24

Weroha SJ, Li SA, Tawfik O, Li JJ (2006) Overexpression of cyclins D1 and D3 during estrogen-induced breast oncogenesis in female ACI rats. Carcinogenesis 27:491–498

Xiong Y, Connolly T, Futcher B, Beach D (1991) Human D-type cyclin. Cell 65:691–699

Zhou H, Kuang J, Zhong L, Kuo WL, Gray JW, Sahin A, Brinkley BR, Sen S (1998) Tumour amplified kinase STK15/BTAK induces centrosome amplification, aneuploidy and transformation. Nature Genet 20:189–193

Androgen-mediated Control of the Cyclin D1-RB Axis: Implications for Prostate Cancer

Karen E. Knudsen[1], *Clay E.S. Comstock*[1], *Nicholas A. Olshavsky*[1], and *Ankur Sharma*[1]

Summary

Prostatic adenocarcinomas are exquisitely dependent on androgen via its cognate receptor (the androgen receptor, AR) for proliferation and survival. This dependence is exploited in the treatment of disseminated disease, wherein ablation of AR activity constitutes first line therapeutic intervention. While initially effective, these strategies ultimately fail, due to inappropriate restoration of AR activity and AR-mediated cellular proliferation. Resultant studies revealed that AR governs the cyclin D1-RB axis, in addition to other phases of the cell cycle. Strikingly, these studies have revealed unexpected cross talk between the AR and several elements of the cell cycle machinery, and aberrations in these pathways have been associated with disease progression. In this review, the molecular communication between AR and the cyclin D1-RB axis will be discussed, with an emphasis on the implications of these pathways for prostate cancer progression and management.

Introduction

Prostatic adenocarcinoma is the most frequently diagnosed malignancy and second leading cause of cancer death amongst men in western countries (Jemal et al. 2005). Significant morbidity associated with the disease results from the failure to effectively manage disseminated prostate cancer, and efforts to improve therapeutic intervention have revealed a pivotal role for hormone action. Local disease can be definitively treated by surgical resection or through radiation therapy, with excellent cure rates for patients presenting with early stage tumors (Kolvenbag and Nash 1999; Nyman et al. 2005). However, late stage and metastatic disease presents a clinical and therapeutic challenge; these tumors respond poorly to standard cytotoxic regimens that act through genomic insult, and lack of effectiveness has been attributed to the relatively indolent nature of the tumor type. Therefore, prostate cancers are treated based on a unique characteristic, in that they are exquisitely dependent on androgen for development, growth, and survival.

The pioneering work of Huggins and Hodges first established that prostate cancers are dependent on serum androgen. Using canine models, these investigators showed that castration of the animals resulted in both an involution of the normal prostate and ablation of spontaneous prostatic adenocarcinomas (Huggins and Hodges 1972).

[1] Department of Cell Biology, ML0521, University of Cincinnati College of Medicine, 3125 Eden Ave., Vontz Center, Cincinnati OH 45267 0521, USA, *e-mail*: karen.knudsen@uc.edu

Melmed et al.
Hormonal Control of Cell Cycle
© Springer-Verlag Berlin Heidelberg 2008

Subsequent studies in cell culture and animal models revealed that androgen ablation triggers cell death or cell cycle arrest of prostate cancer cells (Isaacs 1984; Knudsen et al. 1998). Thus, androgen ablation remains the primary course of treatment for all patients with metastatic (including micrometastatic) disease (Jenster 1999). These therapies are initially effective in the vast majority of patients and result in disease remission. However, recurrent tumors arise within a median of two to three years, wherein androgen signaling has been inappropriately restored (Feldman and Feldman 2001). At present, few therapeutic regimens have been described to significantly manage recurrent prostate cancers, and this is considered an incurable stage of the disease. Thus, androgen action underlies both tumor development and tumor progression in prostatic adenocarcinoma. Given the clear addiction of prostate cancer cells to the androgen signaling axis, a concerted effort has been undertaken to determine the mechanism(s) by which androgen induces prostate cancer cell proliferation and survival.

AR is a Master Regulator of Prostate Cancer Growth and Recurrence

Androgen exerts its biological effects through the androgen receptor (AR), a member of the nuclear receptor superfamily that acts as a ligand dependent transcription factor (Fig. 1; Lee et al. 1995; Trapman and Brinkmann 1996). Testosterone is the most abundant androgen in the sera, but in the prostate it is converted to a more potent androgen, dihydrotestosterone (DHT), through the action of a resident enzyme, 5α-reductase (Russell et al. 1994; Russell and Wilson 1994). Prior to ligand binding, the androgen receptor is held inactive through association with heat shock proteins and is precluded from DNA binding. Ligand binding releases the inhibitory heat shock proteins, and the receptor rapidly translocates to the nucleus, where it binds DNA as a homodimer on androgen responsive elements (AREs) within the regulatory regions of target genes (Trapman and Brinkmann 1996). Furthermore, recruitment of co-activators (which contain or recruit histone acetylases) and chromatin remodeling complexes facilitate transcriptional initiation, and AR-dependent gene expression ensues (Gnanapragasam et al. 2000). The specific combinations of cofactors recruited to AREs provide a mechanism for tissue-specific and ligand-specific gene expression. Through these actions, the AR promotes prostate cancer survival and proliferation (Feldman and Feldman 2001; Trapman and Brinkmann 1996). While the comprehensive cohort of AR target genes that underlie each outcome has yet to be clearly defined, discovery of at least one major AR-dependent target gene, prostate specific antigen (PSA; Stephan et al. 2002), has had a major impact on disease management. Specifically, serum PSA is monitored clinically to detect early stage disease, track tumor burden, monitor the efficacy of therapeutic intervention, and detect the emergence of recurrent tumors post-therapy (Ryan et al. 2006). Thus, readouts of AR activity are critical for the assessment of disease progression and therapeutic outcome.

Disruption of AR action is the major therapeutic goal for management of metastatic disease and can be achieved via multiple mechanisms (Feldman and Feldman 2001; Leewansangtong and Soontrapa 1999). First line treatment ablates AR function through ligand depletion, as achieved through bilateral orchiectomy or through the use of GnRH agonists. Adjuvant or second line therapies involve the use of direct AR antagonists (e.g., bicalutamide) which utilize at least two mechanisms of action (Kolvenbag

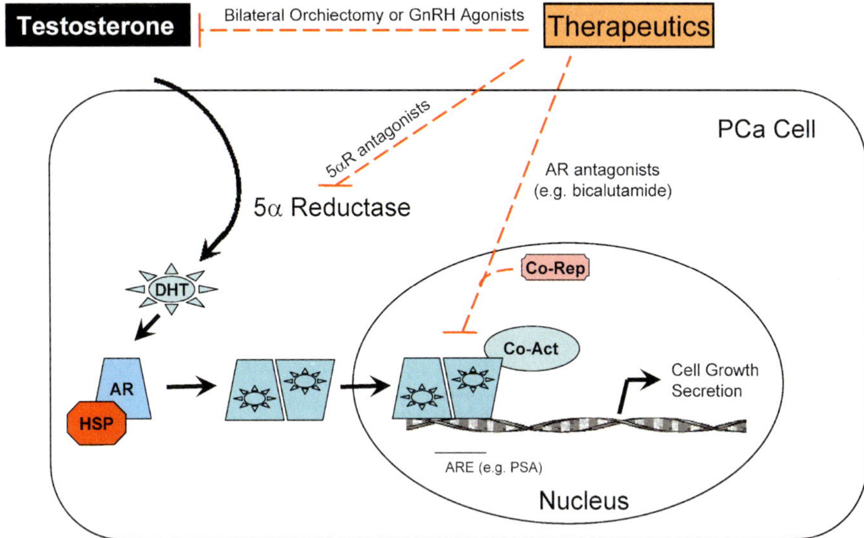

Fig. 1. Androgen signaling and therapeutics in prostate cancer (PCa). Testosterone is converted to a high affinity ligand for the AR, dihydrotestosterone (DHT), in prostate cancer cells. DHT binding causes release of inhibitory heat shock proteins (HSP) from AR and subsequently induces AR homodimer formation, nuclear translocation, and DNA binding to androgen-responsive elements (ARE) of AR target genes (e.g., prostate specific antigen, PSA). Coactivator recruitment (Co-Act) facilitates target gene activation. Disruption of AR activity is the primary treatment for disseminated disease, as achieved by inhibiting androgen synthesis or through the use of direct AR antagonists that compete with DHT for AR binding and recruit corepressors (Co-Rep) to block AR function

and Nash 1999). First, these agents compete for DHT binding. Second, selected AR antagonists trigger the recruitment of transcriptional co-repressors (e.g., NCoR) to AREs, thereby fostering active repression of AR target gene expression (Hodgson et al. 2005). Examination of tumors treated by androgen ablation, with loss of detectable serum PSA, revealed heterogeneous responses concerning cell death or cell cycle arrest amongst dissociated tumor cells. However, this remission is transient, and tumor recurrence is almost invariably observed (Feldman and Feldman 2001; Leewansangtong and Soontrapa 1999). Recurrence is typically preceded by a rise in PSA (also called "biochemical recurrence"; Feldman and Feldman 2001; Trapman and Brinkmann 1996; Visakorpi et al. 1995), and this observation yielded some of the first evidence that tumor progression is associated with restored AR function, despite sustained androgen ablation and/or the use of AR antagonists. Indeed, it is now well established that such "androgen-independent" prostate cancer remains strongly dependent on AR function and that AR activity has been aberrantly restored in recurrent tumors (Chen et al. 2004; Cheng et al. 2006).

Restoration of AR function in recurrent tumors is known to occur through multiple mechanisms (Fig. 2) and in models of cancer is itself causative to resume tumor cell proliferation and therapeutic relapse. First, AR function can be restored through excessive AR expression (including amplification of the locus), as occurs in approximately

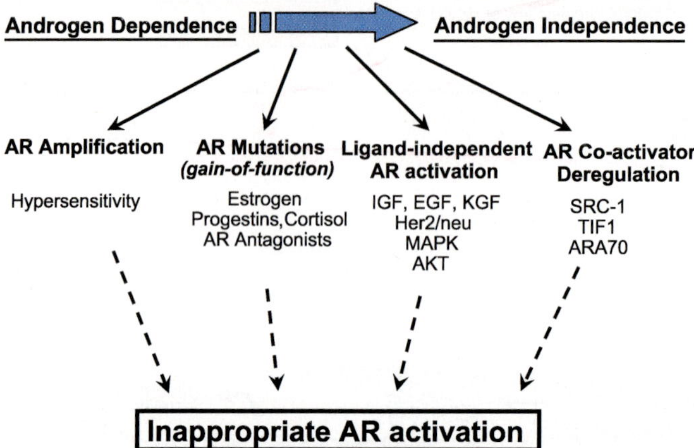

Fig. 2. Mechanisms of therapeutic resistance. Inappropriate activation of AR drives resistance to hormone therapy (androgen independence). This is attributed to multiple pathways, including amplification and mutation of AR, growth factor stimulation, and overexpression of co-activators

30% of recurrent tumors (Chen et al. 2004; Koivisto et al. 1997; Visakorpi et al. 1995). Second, excessive production of specific AR co-activators is observed (e.g., SRC1, TIF1, and ARA70), which can sensitize the receptor to a low ligand environment and/or nullify the effects of AR antagonists (Agoulnik et al. 2005; Gregory et al. 2001; Yeh et al. 1999a). Third, approximately 8–25% of recurrent tumors harbor somatic, gain-of-function mutations of the AR, which render the receptor amenable to activation by a broad spectrum of ligands, including estrogen, progesterone, cortisol, or even some AR antagonists utilized in therapy (Culig et al. 1993; Taplin and Balk 2004). The first AR mutation (T877A) described occurs in the coding region of the ligand binding domain and was identified from patients whose tumors showed a proliferative response to flu-tamide (Veldscholte et al. 1992). Subsequent studies showed that the T877A mutant can use flutamide as an agonist rather than an antagonist and underlies the proliferative response to this antagonist (Masiello et al. 2004). To date, over 600 different mutations of AR have been described, and further studies have shown that some of these mutants may also be altered in their requirement for AR cofactors, thus further facilitating AR activity (http://androgendb.mcgill.ca/). Fourth, AR can be indirectly activated by other signal transduction pathways commonly deregulated in cancer, including MAPK and AKT, although the precise mechanisms underlying these events remain incompletely understood (Gao et al. 2006; Yeh et al. 1999b). Lastly, provocative new data have shown that macrophage invasion into the tumor microenvironment can induce an IL-1ß-dependent signal transduction cascade that disrupts formation of transcriptional repressor complexes initiated by AR antagonists, thus converting the antagonist into an agonist (Zhu et al. 2006).

Combined, these observations support the current hypothesis that AR is a master regulator of prostate cancer cell proliferation and that androgen ablation/antagonists regimens induce an environment of selective pressure to restore AR function. Given the importance of AR as a key determinant of prostate cancer growth and progression, it is imperative to dissect the mechanisms by which AR governs cellular proliferation in prostate cancer cells.

AR Governs the Cyclin D-RB Axis in Prostate Cancer Cells

Analyses of AR-dependent cell cycle progression in prostate cancer cells have shown that androgen is a critical regulator of the G1-S transition (Fig. 3). Prostate cancer cells deprived of androgen arrest in early G1 phase, concomitant with loss of cyclin D1 and cyclin D3 expression, attenuated CDK4 activity (expression unchanged), and hypophosphorylated/actived retinoblastoma tumor suppressor (RB; Knudsen et al. 1998; Xu et al. 2006). Recent studies revealed that androgen induces D-type cyclin expression via mTOR-dependent enhancement of translation (Xu et al. 2006). The ability of androgen to modulate cyclin D translation is distinct from mechanisms utilized by other hormones. For example, estrogen induces cyclin D1 transcription in breast cancer cells, through the ability of its cognate receptor (the estrogen receptor, ER) to directly modulate cyclin D1 regulatory regions (Eeckhoute et al. 2006; Sabbah et al. 1999). Thus, androgen regulation of early G1 events is specific to this class of hormone.

In contrast to the D-type cyclins, cyclin E levels remain unchanged by androgen withdrawal, indicating that alteration of cyclin E expression is not a major mechanism of androgen action. However, cyclin A levels and overall CDK2 activity are diminished upon androgen ablation. These data are consistent with the observation that androgen depletion invokes RB activity, as cyclin A is a well-established target of RB-mediated transcriptional repression. Furthermore, androgen depletion induces p27^{Kip1}, which is likely to contribute to the observed reductions in CDK2 activity (Knudsen et al.

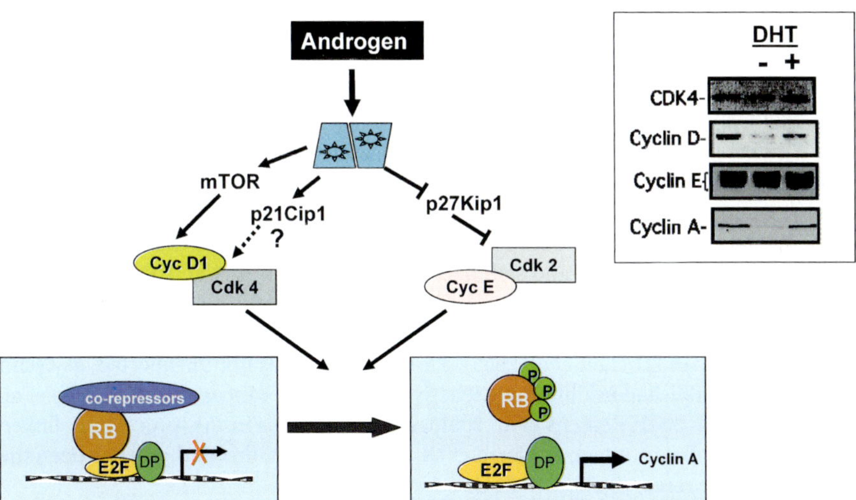

Fig. 3. Androgen-dependent regulation of the G1-S transition. Androgen-mediated induction of G1-S control. Inset data originally appeared in Knudsen et al. (1998)

1998). This supposition is consistent with more recent findings demonstrating that low p27^{Kip1} expression is predictive for shorter time to disease recurrence in prostate cancer (Halvorsen et al. 2003). Similarly, heterozygous PTEN mouse models of prostate cancer have p27^{Kip1} loss, which promotes a tumorigenic phenotype (Gao et al. 2004). Interestingly, upon re-stimulation with androgen, p27^{Kip1} is degraded (Ye et al. 1999). By contrast, p21^{Cip1} expression is lost upon androgen ablation in prostate cancer cells in vitro, which correlates with a higher proliferative index in human tumor specimens (Knudsen et al. 1998; Kolar et al. 2000). Thus, p21^{Cip1} correlates with androgen stimulation and mitogenic proliferation in prostate cancer. Remarkably, p21^{Cip1} has been validated as a direct AR target gene (Lu et al. 1999), and its induction upon androgen ablation may assist in assembling active CDK4/cyclin D1 complexes (Barnes-Ellerbe et al. 2004). In summary, these data culminate in a model wherein androgen induces cyclin D1 accumulation through mTOR, promotes active CDK4/cyclin D1 assembly through p21^{Cip1} induction, and facilitates CDK2 activation through degradation of p27^{Kip1}. These collective events result in RB phosphorylation, de-repression of cyclin A expression, and S-phase progression. Based on this knowledge of AR function, it could be hypothesized that aberrations in the cyclin D-RB axis in cancer could supplant the requirement for androgen and contribute to disease progression. Investigations challenging this hypothesis have revealed novel roles for the D-type cyclins in prostate cancer and a critical function for RB in controlling the response to androgen ablation therapy.

Unique Roles of D-type Cyclins in Prostate Cancer

As described above, the AR uses distinct mechanisms to govern G1-S progression. However, a multitude of studies have demonstrated that there is crosstalk between the two pathways, wherein the cell cycle machinery feeds back on AR to control its action (Fig. 4). The concept that AR is regulated as a function of the cell cycle has been documented (Martinez and Danielsen 2002), and elements of both the G1 and G2/M machinery have been implicated in controlling AR function (Chen et al. 2006; Litvinov et al. 2006). It has recently been shown that AR is degraded in mitosis, and it is suggested that AR may therefore serve as a potential "licensing factor" for prostate cancer cells (Litvinov et al. 2006). However, this remains a loose hypothesis and the evidence for licensing action has not been rigorously addressed. More concrete evidence of cell cycle regulation comes from recent studies wherein it was shown that CDK1 activity fosters AR phosphorylation and stabilizes the receptor (Chen et al. 2006), although it is not clear whether this CDK action is direct. CDK6 has also been implicated as an activator of AR; this function is strikingly independent of its kinase activity and is inhibited by cyclin D1 (Lim et al. 2005). This observation is not unexpected, as cyclin D1 is a well-established inhibitor of AR activity in prostate cancer cells (Knudsen et al. 1999; Petre et al. 2002; Reutens et al. 2001), and aberrations in this process are linked to significant cellular outcomes (Burd et al. 2006). As such, this pathway has been the focus of intense research and will be discussed in detail.

Previous studies have clearly demonstrated that androgen stimulates cyclin D1 accumulation and concomitant CDK4 activation (Knudsen et al. 1998; Xu et al. 2006). However, restoration of cyclin D1 expression under conditions of androgen ablation is

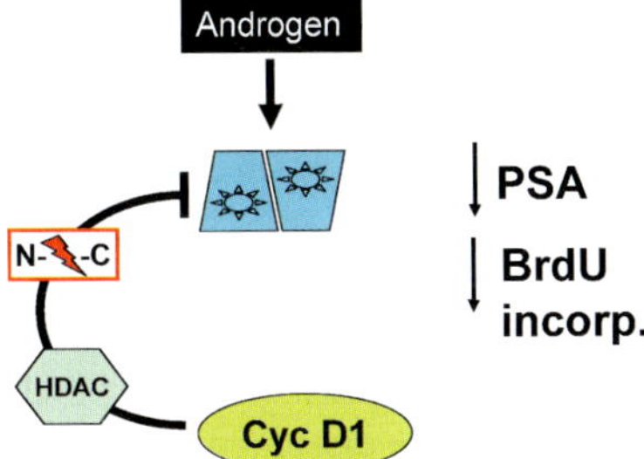

Fig. 4. Cyclin D1-AR negative feedback loop. Stimulation of AR results in accumulation of cyclin D1 leading to CDK4 activation and cell cycle progression. Accumulated cyclin D1 attenuates AR activity by blocking N-C interactions necessary for AR function or by recruitment of histone deacetylase 3 (HDAC3). These events result in PSA expression and attenuated androgen-dependent proliferation. Thus, cyclin D1 modulates the strength and duration of the androgen response

insufficient to drive androgen-independent proliferation (Fribourg et al. 2000). Moreover, it was observed that modest elevations of cyclin D1 in the presence of androgen markedly inhibit (rather than enhance) cellular proliferation (Burd et al. 2005; Petre-Draviam et al. 2003). This unexpected capacity of cyclin D1 to attenuate cell cycle progression is specific to AR-positive, androgen-dependent prostate cancer cells, thus suggesting a putative relationship between cyclin D1 and AR function. Detailed examination of this interaction revealed an unexpected and unique role of cyclin D1 in control of AR activity.

In addition to its ability to modulate CDK4 kinase activity, increasing evidence has demonstrated that cyclin D1 harbors CDK-independent functions in controlling transcription factor action (Coqueret 2002). Cyclin D1 has been shown to directly interact with and modulate a large number of transcription factors, including v-Myb, DMP1, Sp-1, and MyoD. However, the largest class of cyclin D1-associated transcription factors belongs to the nuclear receptor superfamily, including estrogen receptor (ERα), hyroid hormone receptor (TR), PPARγ and AR (Coqueret 2002; Ewen and Lamb 2004). In the case of AR, cyclin D1 binds directly to the N-terminus of the receptor and blocks conformational changes that are required for maximal AR activity upon ligand activation (N-C interaction; Burd et al. 2005; Petre-Draviam et al. 2005). Moreover, cyclin D1 associates with histone deacetylase 3 (HDAC3), and recruitment of HDAC activity is essential for its co-repressor functions (Lin et al. 2002; Petre-Draviam et al. 2005). These actions of cyclin D1 are independent of CDK activity, and a repressor domain within the protein (encoded by amino acids 142-253) has been identified that is capable of supporting both cyclin D1 co-repressor functions (Petre-Draviam et al. 2005). The biological consequence of this event is evident, in that even modest induction of cyclin D1 levels (at stoichiometric levels with the receptor) is sufficient to suppress both AR activity and androgen-dependent proliferation in AR-positive prostate cancer cells (Petre-Draviam et al. 2003). As expected, AR-negative prostate cancer cells are refractory to the repressor function of cyclin D1 (Burd et al. 2006). These data are consistent with observations that AR activity is highly regulated as a function of the cell cycle, wherein cyclin D1 levels inversely correlate with AR activity (Martinez and Danielsen 2002). Moreover, in a mouse model of prostate cancer, cyclin D1 levels decrease as a function of progression, whereas cyclin E levels are elevated; this observation led to the hypothesis of a putative "cyclin switch" that may occur in prostate cancer progression (Maddison et al. 2004a), although this concept has yet to be validated in human specimens. Based on these collective observations, it is hypothesized that cyclin D1 serves as a "negative feedback switch" to modulate androgen-dependent gene expression and concomitant cellular proliferation, thereby governing the strength and duration of the androgen response. Strikingly, recent analyses indicated that these "balancing" functions of cyclin D1 are disrupted in prostate cancer (Knudsen 2006).

Cyclin D1 Aberrations in Prostate Cancer: Localization and Expression

Given the importance of cyclin D1 in proliferative control and its ability to promote oncogenic transformation (Diehl 2002; Gladden and Diehl 2005; Sherr 1995), several studies have investigated cyclin D1 status in human prostate cancer. Initially, these studies compared benign prostatic hyperplasia (BPH) to tumor tissue, but this approach has become less common with the increasing availability of normal tissue adjacent to tumor. As summarized in Table 1, cyclin D1 is rarely amplified (Bubendorf et al. 1999; Das et al. 2005; El Gedaily et al. 2001; Gumbiner et al. 1999; Linja et al. 2001) and most (but not all) immunohistochemical studies have overwhelming shown that cyclin D1 is elevated in prostate cancer (Aaltomaa et al. 1999; Drobnjak et al. 2000; Han et al. 1998; Kallakury et al. 1997; Kolar et al. 2000; Murphy et al. 2005; Shiraishi et al. 1998; Shukla et al. 2004). However, elucidation of the relevance of cyclin D1 expression in prostate cancer has yet to emerge, in part due to the divergent criteria used to define positive cyclin D1 staining. Furthermore, it has been observed that cyclin D1 may be localized to the cytoplasm in prostate tumors (Aaltomaa et al. 1999; Han et al. 1998; Shiraishi et al. 1998). This observation is not entirely unexpected, as cytoplasmic cyclin D1 staining has been noted in other tumor types (Culhaci et al. 2005; Dhar et al. 1999; Dworakowska et al. 2005; Hibberts et al. 1999; Kuramochi et al. 2006; Palmqvist et al. 1998; Sato et al. 1999; Temmim et al. 2006; Tut et al. 2001). These complexities, once resolved, may help to reach a common conclusion concerning the importance of cyclin D1 in prostate cancer tumorigenesis.

Several studies have concluded that increased cyclin D1 holds no independent prognostic significance (Aaltomaa et al. 1999; Kallakury et al. 1997), but a subset of studies have documented positive associations between cyclin D1 and proliferative features such as Ki-67 (Drobnjak et al. 2000; Murphy et al. 2005) and p21^{Cip1} (Kolar et al. 2000). Furthermore, p21^{Cip1} is an important assembly and nuclear translocation factor for the cyclin D1/CDK4 complex, has been shown to be stimulated by androgen (Knudsen et al. 1998), and is a validated AR target gene (Lu et al. 1999). Interestingly, unique roles for p21^{Cip1} in the cytoplasm have been ascribed (Coqueret 2003), suggesting that a connection between p21^{Cip1}, AR, and cyclin D1 localization may be important for prostate cancer progression. These data imply that more study is required and that cyclin D1 status in conjunction with other clinicopathological variables may have predictive value.

Table 1. Cyclin D1 in Human Prostate Cancer

Study	Tissue Description (n)	Method	Result
Amplification			
Gumbiner et.al., 1999	BPH (15), Primary (93), Lymph Node Metastatic (3)	RT-PCR	Primary (4.3%) amplification
Bubendorf et.al.,1999	BPH (32), Primary (223), Recurrent (54), Metastatic (62)	FISH	Primary (1.2%), Recurrent (7.9%), Metastatic (4.7%) amplification
El Gedaily et.al., 2001	Advanced (27)	CGH	Advanced (3.7%) amplification, 3 gains at 11q13
Linja et.al., 2001	BPH (9), Primary (30), Refractory (12)	qPCR	No amplification
Das et.al., 2005	BPH (33), Primary (46) 6 had Bone Metastases	FISH	No amplification, 13 gains at 11q13
Expression			
Kallakury et.al., 1997	Primary (140), Metastatic (19)	IHC	Primary (22.1%), Metastatic (15.8%)
Shiraishi et.al., 1998	Primary (66)	IHC	Primary (30.3%)
Han et.al., 1998	Primary (50) with normal adjacent	IHC	Primary (30.0%), Normal adjacent (18.0%)
Aaltoma et.al., 1999	Primary (187)	IHC	Primary (71.1%), Normal adjacent weakly positive
Drobnjak et.al., 2000	Primary (86), Bone Metastatic (22)	IHC	Primary (11.6%), Bone Metastatic (68.2%)
Kolar et.al., 2000	Primary (89)	IHC	Increased in primary
Shukla et.al., 2004	BPH (3), Primary (6)	Western	Increased in primary
Murphy et.al., 2005	Normal (40), HGPIN (10), Primary (80), AIPC (10)	IHC	Increased in HGPIN, Primary, and AIPC

Toward this end, we recently assessed the expression of cyclin D1 in human tissue from 38 non-neoplastic, 138 prostate tumors, and three metastatic lymph node specimens (Comstock and Knudsen, in press). We show that while cyclin D1 expression is low or absent in normal tissue, its levels are increased in the majority of localized tumors. Surprisingly, four distinct expression profiles were observed in these tumor sets, wherein the largest fraction of cyclin D1-positive tumors showed cytoplasmic restriction. Expression profiles showed some grade specificity; nuclear cyclin D1 staining emerged almost exclusively in the higher-grade tumors. Additionally, PSA expression was lower in the cyclin D1-positive tumors, indicating that cyclin D1 status may affect expression of serum markers that are dependent on AR activity. The relevance of cyclin D1 status to the proliferative index was also considered, wherein tumors with predominantly cytoplasmic cyclin D1 exhibited the lowest proliferative index, even as compared to cyclin D1 negative tumors.

Lastly, nuclear p21$^{\text{Cip1}}$ status was investigated, and p21$^{\text{Cip1}}$ levels frequently associated with a more proliferative and predominantly nuclear cyclin D1 phenotype. Together, these data indicate that cyclin D1 exhibits unique expression profiles in prostate cancer and that the status and/or localization of cyclin D1 expression may be associated with meaningful changes in tumor marker expression and proliferative indices.

Cyclin D1 Aberrations in Prostate Cancer: G/A870 and Alternative Splicing

While the studies described above indicate a potential role for cyclin D1 dysregulation in prostate cancer, more substantive demonstration of cyclin D1 alterations in this disease have emerged from analyses of cyclin D1 polymorphisms and splice variants. A known polymorphism (G/A870) of the cyclin D1 locus exists and has been potentially associated with increased cancer risk or poor prognosis for a number of cancers, including prostate (Koike et al. 2003; Wang et al. 2003). The G/A870 polymorphism is a silent mutation but alters a splice donor site at the exon 4-intron 4 boundary (Betticher et al. 1995). Although these data have yet to be directly challenged, the A-allele is predicted to decrease splicing efficacy and to predispose for the production of a known alternative transcript, deemed "transcript b" (Knudsen 2006). Recent investigations revealed that multiple factors (in addition to the polymorphism) likely govern transcript b production, including a subset of SWI/SNF chromatin remodeling complexes that utilize the BRM ATPase (Batsche et al. 2006). The factors that promote production of this alternate transcript are of high interest, as studies have shown that transcript b may be elevated in tumors and/or independently predictive of poor outcome (Knudsen et al. 2006; Koike et al. 2003; Wang et al. 2003).

The cellular consequence of transcript b production is profound. As attributed to a premature stop codon within intron 4, transcript b results in a divergent protein product, cyclin D1b, which harbors distinct functions from the full-length protein (Betticher et al. 1995). The C-terminus of cyclin D1b is unique, as the protein lacks the PEST domain (implicated in the process of protein destabilization) and a phosphorylation site Thr286 (which controls nuclear export and subsequent protein turnover; Betticher et al. 1995). Predicted loss of these domains led to the hypothesis that the actions elicited by cyclin D1b may be attributed to its increased stability and inability to be exported from the nucleus (Knudsen 2006). Recent evidence demonstrated that, while cyclin D1b is indeed largely nuclear, it is not intrinsically more stable (Lu et al. 2003; Solomon et al. 2003). Functional assessment of cyclin D1b in proliferative control revealed that this protein is a poor activator of CDK4-dependent RB phosphorylation (Solomon et al. 2003). This finding was unexpected, as the functional domains of cyclin D1 required to bind and activate CDK4 are conserved in cyclin D1b. While it would be expected that this deficiency may compromise the oncogenic potential of cyclin D1b, in fact cyclin D1b is a significantly more powerful oncogene than its full-length counterpart. Specifically, cyclin D1b has an enhanced ability to induce cellular transformation of NIH-3T3 murine fibroblasts (Lu et al. 2003; Solomon et al. 2003). Similarly, cyclin D1-deficient murine embryonic fibroblasts acquired anchorage independence when cyclin D1b, but not full-length cyclin D1, was reintroduced (Holley et al. 2005). These collective observations strongly suggest that cyclin D1b harbors increased oncogenic activity, although the mechanisms have yet to be discerned.

With regard to transcriptional control, cyclin D1b is significantly compromised in its ability to regulate estrogen- and androgen-dependent transcription. The ability of

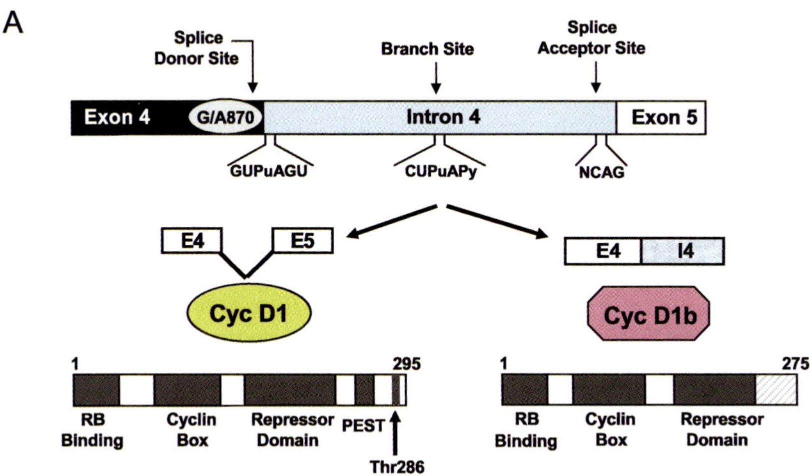

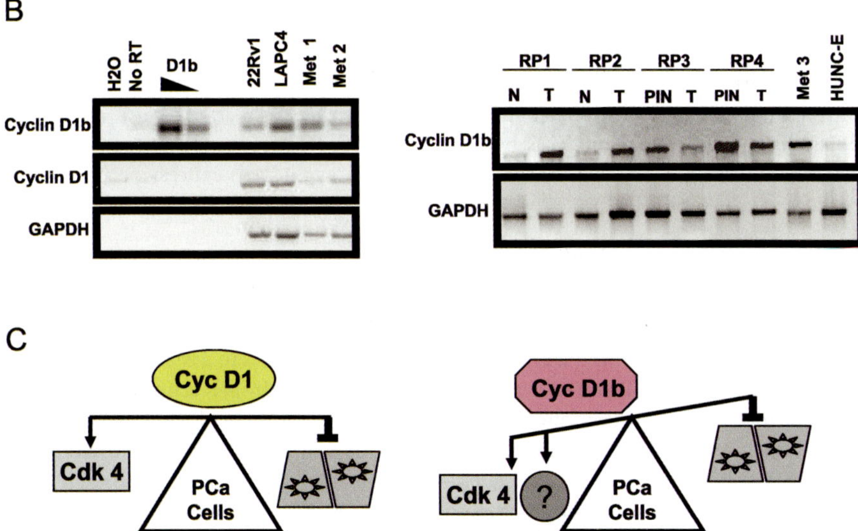

Fig. 5. Cyclin D1b in prostate cancer. (**A**) The cyclin D1b variant arises from a failure to splice at the exon 4/intron 4 boundary. (**B**) The cyclin D1b variant is expressed at high levels in primary and metastatic lesions of prostate cancer. Data extracted from Burd et al. 2006. (**C**) Functional analyses of cyclin D1b show that the protein fails to appropriately regulate AR and yields a proliferative advantage in AR-dependent cells. (Described in Knudsen 2006)

cyclin D1 to modulate the ER is dependent on an "LxxLL" (classical nuclear receptor interaction motif, residues 251 to 255) that is absent in cyclin D1b and therefore has lost the ability to modulate ER activity (Groh and Knudsen, in preparation). The repressor domain of cyclin D1 that is required to bind and modulate AR is mostly intact in cyclin D1b (deletion of amino acids 242-253), but the divergent protein retains the ability to associate with AR both *in vitro* and *in vivo*. However, the cyclin D1b protein is selectively compromised for AR regulation. As demonstrated by monitoring AR activity in transient assays and expression of endogenous AR target genes (e.g., PSA), cyclin D1b is deficient in its ability to regulate AR-dependent transcription. Moreover, cyclin D1b has lost the ability to attenuate androgen-dependent proliferation (unlike full-length cyclin D1); by contrast, this divergent protein promoted cell cycle progression in AR-dependent prostate cancer cells (Burd et al. 2006). Together, these data suggest that cyclin D1b may confer a growth advantage on AR-positive cells by way of its altered ability to modulate AR function (Fig. 5).

The concept that cyclin D1b may facilitate tumor development and/or progression in prostate cancer through a failure to control AR activity is consistent with the established observations that unchecked AR activity is causative for tumor progression (Feldman and Feldman 2001; Trapman and Brinkmann 1996; Visakorpi et al. 1995). Moreover, these data suggest that examination of nuclear cyclin D1 status in tissues should include whether the observed immuno-positivity is attributed to full-length cyclin D1 or the splice variant. This concern may be especially warranted, as recent studies have shown that cyclin D1b is elevated in tumor samples or prostatic intraepithelial neoplasia (PIN) as compared to matched normal tissue from the same individual. Moreover, high cyclin D1b expression was retained in lymph node metastases of prostate cancer (Burd et al. 2006), thus indicating that this presumptive oncogene is likely retained even in late stage disease. Together, these data suggest the intriguing hypothesis that cyclin D1b may facilitate prostate cancer development and/or progression through combinatorial modulation of cell cycle control and androgen-dependent gene expression. This hypothesis is under active scrutiny, and studies to address this premise will clarify the consequence of cyclin D1b expression in prostate cancer.

RB Function in Prostate Cancer and Therapeutic Response

As discussed above, a central cell cycle function of cyclin D1 is to assist in RB inactivation through CDK4/6-dependent phosphorylation, and androgen stimulation utilizes discrete mechanisms to initiate RB inactivation. As such, several models have challenged the impact of RB in the murine prostate. One of the most widely studied rodent models is the *TRAMP* (transgenic adenocarcinoma mouse prostate) transgenic line that utilizes an AR-dependent, prostate specific promoter (probasin) to drive expression of SV40 large T- and small t-antigens in the luminal epithelia. Depending on genetic background, these mice develop high grade PIN and/or prostate cancer within 12 weeks of birth and ultimately develop lung and lymph node metastases by 30 weeks (Gingrich et al. 1997; Greenberg et al. 1995). Androgen deprivation by castration results in decreased tumor incidence as well as the appearance of androgen-independent disease (Gingrich et al. 1997). However, it has been noted that these tumors are typically neuroendocrine in phenotype. A similar mouse model, *LADY*, that expresses only large T-antigen shows less aggressive disease but is also neuroendocrine in phenotype. Neuroendocrine prostate cancers are relatively infrequent in humans; however, they tend to be fairly aggressive and are associated with a poor prognosis (Kasper et al. 1998). Other transgenic lines expressing the SV40 viral oncogenes also develop neuroendocrine-like prostate cancers, driven by the *Cryptdin-2* or Gγ-globin promoters, but these models do not appear to progress through an androgen-dependent stage (Garabedian et al. 1998; Perez-Stable et al. 1997). Thus, while viral oncoproteins that act in part to sequester RB can induce prostate cancer, the utility of these systems for analyzing the consequence of RB loss has been limited.

A more specific challenge of RB action in prostate was elucidated by expressing a mutant of large T-antigen that fails to inactivate p53; this event resulted in PIN lesions followed by focally invasive, well-differentiated adenocarcinomas (Hill et al. 2005). However, these effects are likely attributed to other factors in addition to RB. Tissue recombination studies have further defined the role of RB in prostate cancer progression. Specifically, prostate epithelium from RB-deficient embryonic pelvic viscera, when recombined with wild-type rat urogenital mesenchyme under the kidney capsule of male nude mice, results in hyperplastic disease in 40% of grafted samples (Wang et al. 2000). Similarly, conditional RB deletion in the prostate resulted in focal hyperplasia that is potentially reminiscent of early stage disease (Maddison et al. 2004b). These effects are exacerbated by combinatorial p53 deletion, which results in fast-progressing metastatic carcinomas of the prostate (Zhou et al. 2006). Thus, these data show that inactivation of RB may prime prostate cells to become cancerous when subjected to other insults.

The frequency of RB mutation or deletion in human disease has been investigated. RB has been shown to be lost or inactivated in approximately 30−60% of prostate cancers through disparate mechanisms like point mutations in the coding region of RB gene, deletion in the RB promoter region, loss of heterozygosity (LOH), decreased RB protein expression levels, and loss of p16^{INK4a} (an upstream regulator of RB pathway; Brooks et al. 1995; Ittmann and Wieczorek 1996; Jarrard et al. 2002; Tricoli et al. 1996). Despite the prevalence of these events in prostate cancer, very few studies have addressed the consequence of this event for clinical outcome.

RB is activated upon androgen ablation and, based on cell culture models, this event plays an influential role in the cytostatic response to androgen withdrawal. For example, introduction of viral oncoproteins that act in part to sequester RB function can bypass the androgen requirement in AR-dependent prostate cancer cells (Knudsen et al. 1998). This supposition is supported by a clinical study that observed abnormally low RB mRNA in 36% of patients undergoing combined androgen blockade (Mack et al. 1998). Furthermore, in FISH analysis of genetic aberrations after hormonal therapy using advanced prostate tumor specimens, loss of the RB locus was almost four times more frequent after therapy (Kaltz-Wittmer et al. 2000). Combined, these data indicate that RB inactivation and/or deletion may facilitate the transition to androgen independence.

A recent study challenged this hypothesis *in vitro*, through RNA interference-mediated depletion of RB in AR-dependent prostate cancer cells (Sharma and Knudsen, submitted for publication). These data revealed that, while RB depletion did not confer a proliferative advantage in the presence of androgen, RB-deficient cells failed to elicit a cytostatic response (as compared to RB-positive isogenic controls) when challenged with androgen ablation, AR antagonists, or combined androgen blockade. These data not only indicate that RB ablation can facilitate bypass of first line hormonal therapies but also afford a mechanism to delineate the molecular underpinnings of therapeutic resistance. These studies were extended to determine the impact of RB loss on the response to second line chemotherapeutic intervention, as studies in other cell systems have suggested that loss of RB-dependent DNA damage checkpoints can sensitize cells to cytotoxic agents (Harrington et al. 1998; Knudsen et al. 1998). Indeed, RB-depleted prostate cancer cells demonstrated enhanced susceptibility to cell death induced by a select subset of chemotherapeutic agents (anti-microtubule agents and topoisomerase inhibitor). Combined, these data indicate that RB status may be a critical determinant of therapeutic response in prostate cancer.

Conclusions

The clinical challenges in prostate cancer center on controlling the action of the AR, which is required for both tumor development and disease progression. Selective pressure brought on by androgen ablation typically results in a bypass mechanism to activate the receptor in the absence of ligand and thereby restore AR-dependent cellular proliferation. Thus, dissecting the mechanisms by which AR governs cell cycle progression is instrumental for the design of new strategies to treat recurrent disease. It is apparent that activated AR impinges on the cyclin D1-RB axis to control G1-S progression, and emerging evidence indicates that cross talk between AR and the G1-S machinery serves as an important modulatory node to control the androgen response. Aberrations in these processes can facilitate androgen-independent cellular proliferation and likely contribute to the development of recurrent disease. Future investigations into the consequence of cyclin D1 and RB function in prostate cancer are likely to lead to new avenues of therapeutic intervention.

References

Aaltomaa S, Eskelinen M, Lipponen P (1999) Expression of cyclin A and D proteins in prostate cancer and their relation to clinopathological variables and patient survival. Prostate 38:175–182

Agoulnik IU, Vaid A, Bingman WE 3rd, Erdeme H, Frolov A, Smith CL, Ayala G, Ittmann MM, Weigel NL (2005) Role of SRC-1 in the promotion of prostate cancer cell growth and tumor progression. Cancer Res 65:7959–7967

Barnes-Ellerbe S, Knudsen KE, Puga A (2004) 2,3,7,8-Tetrachlorodibenzo-p-dioxin blocks androgen-dependent cell proliferation of LNCaP cells through modulation of pRB phosphorylation. Mol Pharmacol 66:502–511

Batsche E, Yaniv M, Muchardt C (2006) The human SWI/SNF subunit Brm is a regulator of alternative splicing. Nat Struct Mol Biol 13:22–29

Betticher DC, Thatcher N, Altermatt HJ, Hoban P, Ryder WD, Heighway J (1995) Alternate splicing produces a novel cyclin D1 transcript. Oncogene 11:1005–1011

Brooks JD, Bova GS, Isaacs WB (1995) Allelic loss of the retinoblastoma gene in primary human prostatic adenocarcinomas. Prostate 26:35–39

Bubendorf L, Kononen J, Koivisto P, Schraml P, Moch H, Gasser TC, Willi N, Mihatsch MJ, Sauter G, Kallioniemi OP (1999) Survey of gene amplifications during prostate cancer progression by high-throughout fluorescence in situ hybridization on tissue microarrays. Cancer Res 59:803–806

Burd CJ, Petre CE, Moghadam H, Wilson EM, Knudsen KE (2005) Cyclin D1 binding to the androgen receptor (AR) NH2-terminal domain inhibits activation function 2 association and reveals dual roles for AR corepression. Mol Endocrinol 19:607–620

Burd CJ, Petre CE, Morey LM, Wang Y, Revelo MP, Haiman CA, Lu S, Fenoglio-Preiser CM, Li J, Knudsen ES, Wong J, Knudsen KE (2006) Cyclin D1b variant influences prostate cancer growth through aberrant androgen receptor regulation. Proc Natl Acad Sci USA 103:2190–2195

Chen C, Welsbie DS, Tran C, Baek SH, Chen R, Vessella R, Rosenfeld MG, Sawyers CL (2004) Molecular determinants of resistance to antiandrogen therapy. Nature Med 10:33–39

Chen S, Xu Y, Yuan X, Bubley GJ, Balk SP (2006) Androgen receptor phosphorylation and stabilization in prostate cancer by cyclin-dependent kinase 1. Proc Natl Acad Sci USA 103:15969–15974

Cheng H, Snoek R, Ghaidi F, Cox ME, Rennie PS (2006) Short hairpin RNA knockdown of the androgen receptor attenuates ligand-independent activation and delays tumor progression. Cancer Res 66:10613–10620

Coqueret O (2002) Linking cyclins to transcriptional control. Gene 299:35–55

Coqueret O (2003) New roles for p21 and p27 cell-cycle inhibitors: a function for each cell compartment? Trends Cell Biol:13:65–70

Culhaci N, Sagol O, Karademir S, Astarcioglu H, Astarcioglu I, Soyturk M, Oztop I, Obuz F (2005) Expression of transforming growth factor-beta-1 and p27Kip1 in pancreatic adenocarcinomas: relation with cell-cycle-associated proteins and clinicopathologic characteristics. BMC Cancer 5:98

Culig Z, Hobisch A, Cronauer MV, Cato AC, Hittmair A, Radmayr C, Eberle J, Bartsch G, Klocker H (1993) Mutant androgen receptor detected in an advanced-stage prostatic carcinoma is activated by adrenal androgens and progesterone. Mol Endocrinol 7:1541–1550

Das K, Lau W, Sivaswaren C, Ph T, Fook-Chong S, Sl T, Cheng C (2005) Chromosomal changes in prostate cancer: a fluorescence in situ hybridization study. Clin Genet 68:40–47

Dhar KK, Branigan K, Parkes J, Howells RE, Hand P, Musgrove C, Strange RC, Fryer AA, Redman CW, Hoban PR (1999) Expression and subcellular localization of cyclin D1 protein in epithelial ovarian tumour cells. Brit J Cancer 81:1174–1181

Diehl JA (2002) Cycling to cancer with cyclin D1. Cancer Biol Ther 1:226–231

Drobnjak M, Osman I, Scher HI, Fazzari M, Cordon-Cardo C (2000) Overexpression of cyclin D1 is associated with metastatic prostate cancer to bone. Clin Cancer Res 6:1891–1895

Dworakowska D, Jassem E, Jassem J, Boltze C, Wiedorn KH, Dworakowski R, Skokowski J, Jaskiewicz K, Czestochowska E (2005) Prognostic value of cyclin D1 overexpression in correlation with pRb and p53 status in non-small cell lung cancer (NSCLC). J Cancer Res Clin Oncol 131:479–485

Eeckhoute J, Carroll JS, Geistlinger TR, Torres-Arzayus MI, Brown M (2006) A cell-type-specific transcriptional network required for estrogen regulation of cyclin D1 and cell cycle progression in breast cancer. Genes Dev 20:2513–2526

El Gedaily A, Bubendorf L, Willi N, Fu W, Richter J, Moch H, Mihatsch MJ, Sauter G, Gasser TC (2001) Discovery of new DNA amplification loci in prostate cancer by comparative genomic hybridization. Prostate 46:184–190

Ewen ME, Lamb J (2004) The activities of cyclin D1 that drive tumorigenesis. Trends Mol Med 10:158–162

Feldman BJ, Feldman D (2001) The development of androgen-independent prostate cancer. Nat Rev Cancer 1:34–45

Fribourg AF, Knudsen KE, Strobeck MW, Lindhorst CM, Knudsen ES (2000) Differential requirements for ras and the retinoblastoma tumor suppressor protein in the androgen dependence of prostatic adenocarcinoma cells. Cell Growth Differ 11:361–372

Gao H, Ouyang X, Banach-Petrosky W, Borowsky AD, Lin Y, Kim M, Lee H, Shih WJ, Cardiff RD, Shen MM. Abate-Shen C (2004) A critical role for p27kip1 gene dosage in a mouse model of prostate carcinogenesis. Proc Natl Acad Sci USA 101:17204–17209

Gao H, Ouyang X, Banach-Petrosky WA, Gerald WL, Shen MM, Abate-Shen C (2006) Combinatorial activities of Akt and B-Raf/Erk signaling in a mouse model of androgen-independent prostate cancer. Proc Natl Acad Sci USA 103:14477–14482

Garabedian EM, Humphrey PA, Gordon JI (1998) A transgenic mouse model of metastatic prostate cancer originating from neuroendocrine cells. Proc Natl Acad Sci USA 95:15382–15387

Gingrich JR, Barrios RJ, Kattan MW, Nahm HS, Finegold MJ Greenberg NM (1997) Androgen-independent prostate cancer progression in the TRAMP model. Cancer Res 57:4687–4691

Gladden AB, Diehl JA (2005) Location, location, location: the role of cyclin D1 nuclear localization in cancer. J Cell Biochem 96:906–913

Gnanapragasam VJ, Robson CN, Leung HY, Neal DE (2000) Androgen receptor signalling in the prostate. BJU Int 86:1001–1013

Greenberg NM, DeMayo F, Finegold MJ, Medina D, Tilley WD, Aspinall JO, Cunha GR, Donjacour AA, Matusik RJ, Rosen JM (1995) Prostate cancer in a transgenic mouse. Proc Natl Acad Sci USA 92:3439–3443

Gregory CW, Johnson RT Jr, Mohler JL, French FS, Wilson EM (2001) Androgen receptor stabilization in recurrent prostate cancer is associated with hypersensitivity to low androgen. Cancer Res 61:2892–2898

Gumbiner LM, Gumerlock PH, Mack PC, Chi SG, deVere White RW, Mohler JL, Pretlow TG, Tricoli JV (1999) Overexpression of cyclin D1 is rare in human prostate carcinoma. Prostate 38:40–45

Halvorsen OJ, Haukaas SA, Akslen LA (2003) Combined loss of PTEN and p27 expression is associated with tumor cell proliferation by Ki-67 and increased risk of recurrent disease in localized prostate cancer. Clin Cancer Res 9:1474–1479

Han EK, Lim JT, Arber N, Rubin MA, Xing WQ, Weinstein IB (1998) Cyclin D1 expression in human prostate carcinoma cell lines and primary tumors. Prostate 35:95–101

Harrington EA, Bruce L, Harlow E, Dyson N (1998) pRB plays an essential role in cell cycle arrest induced by DNA damage. Proc Natl Acad Sci USA 95:11945–11950

Hibberts NA, Simpson DJ, Bicknell JE, Broome JC, Hoban PR, Clayton RN, Farrell WE (1999) Analysis of cyclin D1 (CCND1) allelic imbalance and overexpression in sporadic human pituitary tumors. Clin Cancer Res 5:2133–2139

Hill R, Song Y, Cardiff RD, Van Dyke T (2005) Heterogeneous tumor evolution initiated by loss of pRb function in a preclinical prostate cancer model. Cancer Res 65:10243–10254

Hodgson MC, Astapova I, Cheng S, Lee LJ, Verhoeven MC, Choi E, Balk SP, Hollenberg AN (2005) The androgen receptor recruits nuclear receptor CoRepressor (N-CoR) in the presence of mifepristone via its N and C termini revealing a novel molecular mechanism for androgen receptor antagonists. J Biol Chem 280:6511–6519

Holley SL, Heighway J, Hoban PR (2005) Induced expression of human CCND1 alternative transcripts in mouse Cyl-1 knockout fibroblasts highlights functional differences. Int J Cancer 114:364–370

Huggins C, Hodges CV (1972) Studies on prostatic cancer. I. The effect of castration, of estrogen and androgen injection on serum phosphatases in metastatic carcinoma of the prostate. CA Cancer J Clin 22:232–240

Isaacs JT (1984) Antagonistic effect of androgen on prostatic cell death. Prostate 5:545–557

Ittmann MM, Wieczorek R (1996) Alterations of the retinoblastoma gene in clinically localized, stage B prostate adenocarcinomas. Human Pathol 27:28–34

Jarrard DF, Modder J, Fadden P, Fu V, Sebree L, Heisey D, Schwarze SR, Friedl A (2002) Alterations in the p16/pRb cell cycle checkpoint occur commonly in primary and metastatic human prostate cancer. Cancer Lett 185:191–199

Jemal A, Murray T, Ward E, Samuels A, Tiwari RC, Ghafoor A, Feuer EJ, Thun MJ (2005) Cancer statistics, 2005. CA Cancer J Clin 55:10–30

Jenster G (1999) The role of the androgen receptor in the development and progression of prostate cancer. Semin Oncol 26:407–421

Kallakury BV, Sheehan CE, Ambros RA, Fisher HA, Kaufman RP Jr, Ross JS (1997) The prognostic significance of p34cdc2 and cyclin D1 protein expression in prostate adenocarcinoma. Cancer 80:753–763

Kaltz-Wittmer C, Klenk U, Glaessgen A, Aust DE, Diebold J, Lohrs U, Baretton GB (2000) FISH analysis of gene aberrations (MYC, CCND1, ERBB2, RB, and AR) in advanced prostatic carcinomas before and after androgen deprivation therapy. Lab Invest 80:1455–1464

Kasper S, Sheppard PC, Yan Y, Pettigrew N, Borowsky AD, Prins GS, Dodd JG, Duckworth ML, Matusik RJ (1998) Development, progression, and androgen-dependence of prostate tumors in probasin-large T antigen transgenic mice: a model for prostate cancer. Lab Invest 78:319–333

Knudsen KE (2006) The cyclin D1b splice variant: an old oncogene learns new tricks. Cell Div 1:15

Knudsen KE, Arden KC, Cavenee WK (1998) Multiple G1 regulatory elements control the androgen-dependent proliferation of prostatic carcinoma cells. J Biol Chem 273:20213–20222

Knudsen KE, Cavenee WK, Arden KC (1999). D-type cyclins complex with the androgen receptor and inhibit its transcriptional transactivation ability. Cancer Res 59:2297–2301

Knudsen KE, Diehl JA, Haiman CA, Knudsen ES (2006) Cyclin D1: polymorphism, aberrant splicing and cancer risk. Oncogene 25:1620–1628

Koike H, Suzuki K, Satoh T, Ohtake N, Takei T, Nakata S, Yamanaka H (2003) Cyclin D1 gene polymorphism and familial prostate cancer: the AA genotype of A870G polymorphism is associated with prostate cancer risk in men aged 70 years or older and metastatic stage. Anticancer Res 23:4947–4951

Koivisto P, Kononen J, Palmberg C, Tammela T, Hyytinen E, Isola J, Trapman J, Cleutjens K, Noordzij A, Visakorpi T, Kallioniemi OP (1997) Androgen receptor gene amplification: a possible molecular mechanism for androgen deprivation therapy failure in prostate cancer. Cancer Res 57:314–319

Kolar Z, Murray PG, Scott K, Harrison A, Vojtesek B, Dusek J (2000) Relation of Bcl-2 expression to androgen receptor, p21WAF1/CIP1, and cyclin D1 status in prostate cancer. Mol Pathol 53:15–18

Kolvenbag GJ, Nash A (1999) Bicalutamide dosages used in the treatment of prostate cancer. Prostate 39:47–53

Kuramochi J, Arai T, Ikeda S, Kumagai J, Uetake H, Sugihara K (2006) High Pin1 expression is associated with tumor progression in colorectal cancer. J Surg Oncol 94:155–160

Lee C, Sutkowski DM, Sensibar JA, Zelner D, Kim I, Amsel I, Shaw N, Prins GS, Kozlowski JM (1995) Regulation of proliferation and production of prostate-specific antigen in androgen-sensitive prostatic cancer cells, LNCaP, by dihydrotestosterone. Endocrinology 136:796–803

Leewansangtong S, Soontrapa S (1999) Hormonal ablation therapy for metastatic prostatic carcinoma: a review. J Med Assoc Thai 82:192–205

Lim JT, Mansukhani M, Weinstein IB (2005) Cyclin-dependent kinase 6 associates with the androgen receptor and enhances its transcriptional activity in prostate cancer cells. Proc Natl Acad Sci USA 102:5156–5161

Lin HM, Zhao L, Cheng SY (2002) Cyclin D1 Is a Ligand-independent Co-repressor for Thyroid Hormone Receptors. J Biol Chem 277:28733–28741

Linja MJ, Savinainen KJ, Saramaki OR, Tammela TL, Vessella RL, Visakorpi T (2001) Amplification and overexpression of androgen receptor gene in hormone-refractory prostate cancer. Cancer Res 61:3550–3555

Litvinov IV, Vander Griend DJ, Antony L, Dalrymple S, De Marzo AM, Drake CG, Isaacs JT (2006) Androgen receptor as a licensing factor for DNA replication in androgen-sensitive prostate cancer cells. Proc Natl Acad Sci USA 103:15085–15090

Lu F, Gladden AB, Diehl JA (2003) An alternatively spliced cyclin D1 isoform, cyclin D1b, is a nuclear oncogene. Cancer Res 63:7056–7061

Lu S, Liu M, Epner DE, Tsai SY, Tsai MJ (1999) Androgen regulation of the cyclin-dependent kinase inhibitor p21 gene through an androgen response element in the proximal promoter. Mol Endocrinol 13:376–384

Mack PC, Chi SG, Meyers FJ, Stewart SL, deVere White RW, Gumerlock PH (1998) Increased RB1 abnormalities in human primary prostate cancer following combined androgen blockade. Prostate 34:145–151

Maddison LA, Huss WJ, Barrios RM, Greenberg NM (2004a) Differential expression of cell cycle regulatory molecules and evidence for a "cyclin switch" during progression of prostate cancer. Prostate 58:335–344

Maddison LA, Sutherland BW, Barrios RJ, Greenberg NM (2004) Conditional deletion of Rb causes early stage prostate cancer. Cancer Res 64:6018–6025

Martinez ED, Danielsen M (2002b) Loss of androgen receptor transcriptional activity at the G(1)/S transition. J Biol Chem 277:29719–29729

Masiello D, Chen SY, Xu Y, Verhoeven MC, Choi E, Hollenberg AN, Balk SP (2004) Recruitment of beta-catenin by wild-type or mutant androgen receptors correlates with ligand-stimulated growth of prostate cancer cells. Mol Endocrinol 18:2388–2401

Murphy C, McGurk M, Pettigrew J, Santinelli A, Mazzucchelli R, Johnston PG, Montironi R, Waugh DJ (2005) Nonapical and cytoplasmic expression of interleukin-8, CXCR1, and CXCR2 correlates with cell proliferation and microvessel density in prostate cancer. Clin Cancer Res 11:4117–4127

Nyman CR, Andersen JT, Lodding P, Sandin T, Varenhorst E (2005) The patient's choice of androgen-deprivation therapy in locally advanced prostate cancer: bicalutamide, a gonadotrophin-releasing hormone analogue or orchidectomy. BJU Int 96:1014–1018

Palmqvist R, Stenling R, Oberg A, Landberg G (1998) Expression of cyclin D1 and retinoblastoma protein in colorectal cancer. Eur J Cancer 34:1575–1581

Perez-Stable C, Altman NH, Mehta PP, Deftos LJ, Roos BA (1997) Prostate cancer progression, metastasis, and gene expression in transgenic mice. Cancer Res 57:900–906

Petre-Draviam CE, Cook SL, Burd CJ, Marshall TW, Wetherill YB, Knudsen KE (2003) Specificity of cyclin D1 for androgen receptor regulation. Cancer Res 63:4903–4913

Petre-Draviam CE, Williams EB, Burd CJ, Gladden A, Moghadam H, Meller J, Diehl JA, Knudsen KE (2005) A central domain of cyclin D1 mediates nuclear receptor corepressor activity. Oncogene 24:431–444

Petre CE, Wetherill YB, Danielsen M, Knudsen KE (2002) Cyclin D1: mechanism and consequence of androgen receptor co-repressor activity. J Biol Chem 277:2207–2215

Reutens AT, Fu M, Wang C, Albanese C, McPhaul MJ, Sun Z, Balk SP, Janne OA, Palvimo JJ, Pestell RG (2001) Cyclin D1 binds the androgen receptor and regulates hormone-dependent signaling in a p300/CBP-associated factor (P/CAF)-dependent manner. Mol Endocrinol 15:797–811

Russell DW, Wilson JD (1994) Steroid 5 alpha-reductase: two genes/two enzymes. Annu Rev Biochem 63:25–61

Russell DW, Berman DM, Bryant JT, Cala KM, Davis DL, Landrum CP, Prihoda JS, Silver RI, Thigpen AE, Wigley WC (1994) The molecular genetics of steroid 5 alpha-reductases. Recent Prog Horm Res 49:275–284

Ryan CJ, Smith A, Lal P, Satagopan J, Reuter V, Scardino P, Gerald W, Scher HI (2006) Persistent prostate-specific antigen expression after neoadjuvant androgen depletion: an early predictor of relapse or incomplete androgen suppression. Urology 68:834–839

Sabbah M, Courilleau D, Mester J, Redeuilh G (1999) Estrogen induction of the cyclin D1 promoter: involvement of a cAMP response-like element. Proc Natl Acad Sci USA 96:11217–11222

Sato Y, Itoh F, Hareyama M, Satoh M, Hinoda Y, Seto M, Ueda R, Imai K (1999) Association of cyclin D1 expression with factors correlated with tumor progression in human hepatocellular carcinoma. J Gastroenterol 34:486–493

Sherr CJ (1995) D-type cyclins. Trends Biochem Sci 20:187–190

Shiraishi T, Watanabe M, Muneyuki T, Nakayama T, Morita J, Ito H, Kotake T, Yatani R (1998) A clinicopathological study of p53, p21 (WAF1/CIP1) and cyclin D1 expression in human prostate cancers. Urol Int 6:90–94

Shukla S, MacLennan GT, Fu P, Patel J, Marengo SR, Resnick MI, Gupta S (2004) Nuclear factor-kappaB/p65 (Rel A) is constitutively activated in human prostate adenocarcinoma and correlates with disease progression. Neoplasia 6:390–400

Solomon DA, Wang Y, Fox SR, Lambeck TC, Giesting S, Lan Z, Senderowicz AM, Conti CJ, Knudsen ES (2003) Cyclin D1 splice variants. Differential effects on localization, RB phosphorylation, and cellular transformation. J Biol Chem 278:30339–30347

Stephan C, Jung K, Diamandis EP, Rittenhouse HG, Lein M, Loening SA (2002) Prostate-specific antigen, its molecular forms, and other kallikrein markers for detection of prostate cancer. Urology 59:2–8

Taplin ME, Balk SP (2004) Androgen receptor: a key molecule in the progression of prostate cancer to hormone independence. J Cell Biochem 91:483–490

Temmim L, Ebraheem AK, Baker H, Sinowatz F (2006) Cyclin D1 protein expression in human thyroid gland and thyroid cancer. Anat Histol Embryol 35:125–129

Trapman J, Brinkmann AO (1996) The androgen receptor in prostate cancer. Pathol Res Pract 192:752–760

Tricoli JV, Gumerlock PH, Yao JL, Chi SG, D'Souza SA, Nestok BR, deVere White RW (1996) Alterations of the retinoblastoma gene in human prostate adenocarcinoma. Genes Chromosomes Cancer 15:108–114

Tut VM, Braithwaite KL, Angus B. Neal DE, Lunec J, Mellon JK (2001) Cyclin D1 expression in transitional cell carcinoma of the bladder: correlation with p53, waf1, pRb and Ki67. Br J Cancer 84:270–275

Veldscholte J, Berrevoets CA, Ris-Stalpers C, Kuiper GG, Jenster G, Trapman J, Brinkmann AO, Mulder E (1992) The androgen receptor in LNCaP cells contains a mutation in the ligand binding domain which affects steroid binding characteristics and response to antiandrogens. J Steroid Biochem Mol Biol 41:665–669

Visakorpi T, Hyytinen E, Koivisto P, Tanner M, Keinanen R, Palmberg C, Palotie A, Tammela T, Isola J, Kallioniemi OP (1995) In vivo amplification of the androgen receptor gene and progression of human prostate cancer. Nature Genet 9:401–406

Wang L, Habuchi T, Mitsumori K, Li Z, Kamoto T, Kinoshita H, Tsuchiya N, Sato K, Ohyama C, Nakamura A, Ogawa O, Kato T (2003) Increased risk of prostate cancer associated with AA genotype of cyclin D1 gene A870G polymorphism. Int J Cancer 103:116–120

Wang Y, Hayward SW, Donjacour AA, Young P, Jacks T, Sage J, Dahiya R, Cardiff RD, Day ML, Cunha GR (2000) Sex hormone-induced carcinogenesis in Rb-deficient prostate tissue. Cancer Res 60:6008–6017

Xu Y, Chen SY, Ross KN, Balk SP (2006) Androgens induce prostate cancer cell proliferation through mammalian target of rapamycin activation and post-transcriptional increases in cyclin D proteins. Cancer Res 66:7783–7792

Ye D, Mendelsohn J, Fan Z (1999) Androgen and epidermal growth factor down-regulate cyclin-dependent kinase inhibitor p27Kip1 and costimulate proliferation of MDA PCa 2a and MDA PCa 2b prostate cancer cells. Clin Cancer Res 5:2171–2177

Yeh S, Chang HC, Miyamoto H, Takatera H, Rahman M, Kang HY, Thin TH, Lin HK, Chang C (1999a) Differential induction of the androgen receptor transcriptional activity by selective androgen receptor coactivators. Keio J Med 48:87–92

Yeh S, Lin HK, Kang HY, Thin TH, Lin MF, Chang C (1999b) From HER2/Neu signal cascade to androgen receptor and its coactivators: a novel pathway by induction of androgen target genes through MAP kinase in prostate cancer cells. Proc Natl Acad Sci USA 96:5458–5463

Zhou Z, Flesken-Nikitin A, Corney DC, Wang W, Goodrich DW, Roy-Burman P, Nikitin AY (2006) Synergy of p53 and Rb Deficiency in a Conditional Mouse Model for Metastatic Prostate Cancer. Cancer Res 66:7889–7898

Zhu P, Baek SH, Bourk EM, Ohgi KA, Garcia-Bassets I, Sanjo H, Akira S, Kotol PF, Glass CK, Rosenfeld MG, Rose DW (2006) Macrophage/cancer cell interactions mediate hormone resistance by a nuclear receptor derepression pathway. Cell 124:615–629

Pituitary Trophic Status as a Tumorigenic Determinant

Vera Chesnokova[1], Run Yu[1], Anat Ben-Shlomo[1], and *Shlomo Melmed[1]*

Pituitary tumors are commonly encountered, benign adenomas that arise from cells of the anterior pituitary gland. These tumors arise from highly differentiated cells expressing unique hormone gene products, including growth hormone, prolactin, ACTH, TSH, FSH and LH. They may be functional and actively secrete hormones, leading to characteristic clinical features including acromegaly, Cushing's disease, features of hyperprolactinemia or, rarely, hyperthyroidism. Commonly, they are nonfunctional, leading primarily to hypogonadism and compressive pituitary failure. These monoclonal neoplasms account for $\sim 15\%$ of all intracranial tumors, and malignant transformation very rarely occurs (Melmed 2003). The direct pathogenesis of these tumors remains elusive, and genetic, cellular, animal and human pathologic models have been developed to further understand pituitary tumor origins. Overall, the body of reported work indicates that the etiology of these tumors reflects disordered hypothalamic hormone action, intrapituitary growth factor or oncogene dysfunction, or aberrant peripheral hormone feedback control. Several lines of evidence support the notion that pituitary cell trophic status – i.e., hypotrophic, normal or hypertrophic – is an important determinant of pituitary adenoma development. Understanding mechanisms subserving pituitary cell proliferation provides insight into the unique pathogenesis of pituitary growth disorders, including tumors, hyperplasia, hypoplasia, and genetic disorders of pituitary function – all of which are associated with aberrant pituitary cell growth.

Mature, hormone-secreting anterior pituitary cells do not replicate appreciably and exhibit slow turnover rates. Subsequent postnatal alterations in pituitary size are determined throughout the life span by both extrinsic and intrinsic factors. Pituitary hyperplasia may be caused by several factors, including loss of negative feedback by peripheral organ failure (e.g., thyroid adrenal or gonad), pregnancy, estrogen treatment and hypothalamic hormone excess. The latter is exemplified by carcinoid (and other neuroendocrine tumors) elaborating ectopic GHRH, which stimulates pituitary somatotroph cell growth and growth hormone hypersecretion. Intriguingly, pituitary tumors do not arise with appreciably higher frequency during pregnancy or estrogen treatment, despite significant pituitary gland expansion under these conditions. There is, however, the plausible theory that discrete pituitary adenomas arise in focal areas of pituitary hyperplasia that are no longer evident by the time the adenoma is resected and available for pathologic assessment. In fact, normal pituitary tissue surrounding pituitary adenomas is usually not hyperplastic. In contrast to humans, pituitary adenomas

---

[1] Cedars-Sinai Medical Center 8700 Beverly Blvd., Room 2015 Los Angeles, CA 90048

Address to correspondence: Shlomo Melmed, M.D. Cedars-Sinai Medical Center 8700 Beverly Blvd., Room 2015 Los Angeles, CA 90048, *e-mail*: shlomo.melmed@cshs.org

Melmed et al.
Hormonal Control of Cell Cycle
© Springer-Verlag Berlin Heidelberg 2008

in mice are often preceded by pituitary hyperplasia. Thus, upregulated pituitary-driven growth factors or oncogenes result in well documented pituitary hyperplasia prior to discrete adenoma development in transgenic mice. Pituitary hypoplasia is encountered in mice and humans with developmental defects, including transcription factor mutations (e.g. Pit-1, Prop-1), and pituitary mass generally shrinks with age. Hereditary or sporadic mutations leading to pituitary hypoplasia may lead to structural defects, with commonly associated anterior pituitary hormone deficits and clinical features of growth, thyroid, adrenal, and or reproductive failure. Alternatively, pituitary hypoplasia may be associated with no appreciable pituitary hormone deficiency.

Uniquely, pituitary carcinomas are extremely rare, and the commonly seen pituitary neoplasms are invariably benign. In discrete, wellcircumscribed pituitary adenomas, mitotic activity is relatively low even in aggressively growing tumors. True extrapituitary distant metastases have only been documented in a very limited number of reports worldwide. Therefore, a diagnosis of pituitary adenoma is highly unlikely to be associated with malignancy. This finding contrasts with tumors arising from more rapidly replicating tissues, such as the gastro-intestinal tract, in which the cascade of benign to malignant transformation is not uncommonly encountered.

Senescence, or proliferative arrest, may be due to aging or to replicative arrest in response to external stresses, including UV radiation or hypoxia. The latter form of senescence is mediated by oncogene pathways. Cellular senescence is characterized by activation of the ARF/p53/p21 or Rb/p16 senescence pathways, as well as up-regulation of cell cycle progression inhibitors, including p19^{ARF} (p19), p21^{Cip1} (p21) and p16^{INK4A} (p16). As in vivo senescent markers are preferentially expressed in benign but not in malignant adenocarcinomas, we hypothesize that pituitary senescence accounts for the overwhelming predominance of benign vs. malignant pituitary tumors. We present evidence favoring this hypothesis derived from the *Pttg*-null mouse.

PTTG is a mammalian securin protein that restricts separase activity in separating sister chromatids during mitosis. PTTG acts to inhibit separase, which cleaves sister chromatids with fidelity during mitosis (Zhou et al. 1999). PTTG is overexpressed in several endocrine (pituitary, thyroid, breast) tumors, as well as in colon cancer. PTTG is also a powerful transforming gene, regulates angiogenesis, and acts as a transcriptional activator. Overall genomic stability requires intact PTTG function and appropriate, cell cycle-related intracellular stability and degradation. PTTG1 abundance correlates with tumor invasiveness and size, and the protein is induced during early phases of estrogen-induced rat prolactinoma development. PTTG is reported to be a prognostic indicator for thyroid cancer, lymph node invasion and breast cancer recurrence, and for colorectal cancer invasiveness (Heaney et al. 1999). PTTG1 transcriptional actions include binding to p53, inhibiting p53 transcriptional activity and interacting with Ku, the regulatory subunit of DNA-dependent protein kinase, activating c-Myc and inducing bFGF.

How in fact PTTG1 overexpression mediates tumorigenesis, and the functional mechanisms underlying cellular PTTG1 action are still unclear.

Pituitary-directed transgenic *Pttg* over-expression results in focal pituitary hyperplasia and adenoma formation. When driven by the pituitary specific α-glycoprotein subunit (αGSU) promoter, overexpressed transgenic PTTG causes LH, GH and TSH-secreting adenomas with respective trophic hormone hypersecretion (Abbud et al. 2005). High LH levels lead to elevated testosterone, and transgenic mice harboring

GH-secreting adenomas express high IGF-I levels. These dysfunctions are associated with profound prostatic and seminal vesicle hypertrophy. Mice lacking *Pttg* exhibit pituitary and pancreatic hypoplasia. These mice develop male-selective, insulin-secreting islet β cell proliferation arrest, markedly decreased β cell mass and turnover, hypo-insulinemia, and ultimately hyperglycemia (Wang et al. 2003). These animals exhibit no evidence of auto-immune infiltrative pancreatic processes; neither do they show peripheral insulin signaling defects. The tissue-selective impact of *Pttg* deletion, i.e., pituitary, pancreatic, and splenic hypoplasia, in the face of apparently intact early post-natal functional differentiation of these glands, is intriguing. *Pttg* is therefore required for cell replication in "slow turnover" tissues. Furthermore, redundant securin systems are likely operative to compensate for *Pttg* loss in more rapidly replicative tissues. MEFs derived from these animals also exhibit triradial and quadriradial chromosomes and premature centromere separation. These latter findings are consistent with aneuploidy and support the critical role of PTTG in assuring that sister chromatid separation occurs with fidelity. The pituitary glands in these mice are hypoplastic, with diminished BrdU incorporation, low Ki67 expression levels, and low weight. *Pttg* deletion leads to pituitary $p21^{Cip1}$ induction with slowing of young pituitary gland proliferation. $Pttg^{-/-}$ mice were therefore crossbred with $Rb^{+/-}$ mice, which usually develop pituitary tumors at high penetrance. The resultant compound $Pttg^{-/-}Rb^{+/-}$ transgenic knockout animals are apparently protected from pituitary tumor formation. $Rb^{+/-}$ mice develop pituitary tumors with almost 100% penetrance, whereas only 30% of doubly mutant $Rb^{+/-}$ mice with deleted *Pttg* develop later pituitary tumors (Chesnokova et al. 2005).

The ARF/p53/p21 senescence pathway (Campisi 2005) is activated in the $Pttg^{-/-}$ pituitary gland, and pituitary anti-apoptotic Bcl-2 protein levels were elevated in both $Pttg^{-/-}$ and doubly mutant $Rb^{+/-}Pttg^{-/-}$ animals, consistent with senescent features. *Pttg* deletion results in reduced RB phosphorylation at Ser807/811, a residue preferentially phosphorylated by cyclin E-Cdk2 complexes. Overall, the senescent changes observed in *Pttg*-deficient pituitary glands lend support to the requirement of PTTG for mature rather than embryonic pituitary cell cycle progression and proliferation. The observed senescent changes are consistent with the small pituitary size and decreased $Pttg^{-/-}$ pituitary cell proliferation. As *Pttg*-dependent pituitary senescence is not associated with telomerase dysfunction, the observed pituitary senescence pathway is likely not characterized by premature aging. Senescence can be triggered by aneuploidy or DNA damage, and *Pttg* deficiency triggers aneuploid features and DNA damage-signaling genes are also activated. We surveyed PTTG1 effects on 20,000 gene promoters by ChIP-on-Chip assay and showed that PTTG1 acts as a global transcription factor. PTTG1 regulates the G1/S transition by coordinating with Sp1, by directly interacting with cell cycle regulating proteins, including p21 and CCND3. These functions may underlie the transforming activity of PTTG.

ChiP and EMSA assays performed in AtT20 mouse pituitary corticotrophs expressing high *Pttg* levels showed that PTTG was recruited to the p21 promoter region, (1–120 nt upstream from the transcription start site). Endogenous Sp1 also associated with PTTG, suggesting that PTTG regulates p21 promoter activity by binding to an Sp1-binding site. Results of ChIP-on-Chip analysis showed that PTTG1 bound to 746 gene promoters and the transcriptional pattern analysis showed that PTTG1 interacted with Sp1 (p < 0.000001). These findings were confirmed by co-immunoprecipitation and His-pull down assays and EMSA, which mapped the interaction domains between

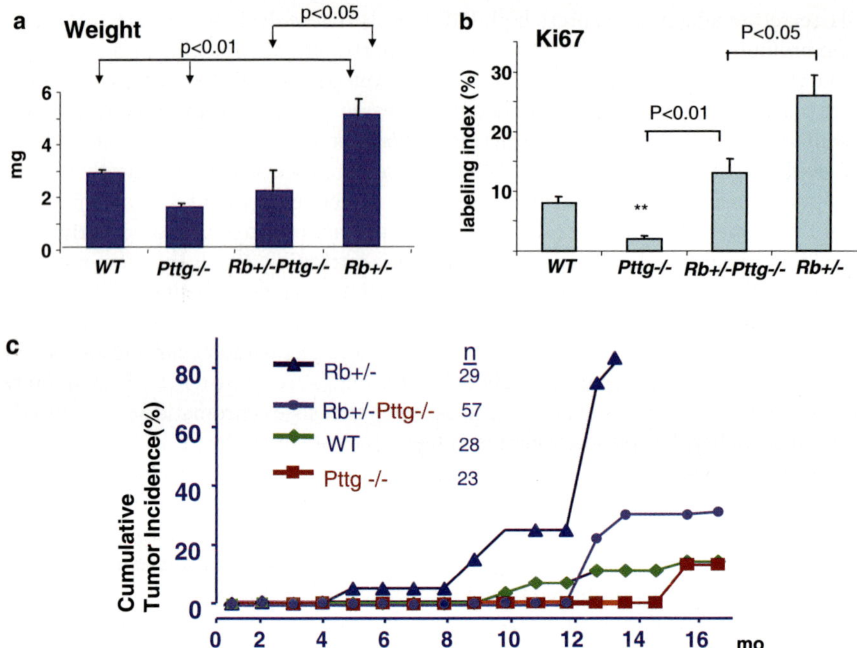

Fig. 1. *Pttg* deletion decreases pituitary weight, cell proliferation and pituitary tumor incidence in *Rb+/-* mice. (**a**) Pituitary weight of 2-month-old pre-tumorous $Rb^{+/+}Pttg^{+/+}$, $Rb^{+/-}Pttg^{+/+}$, $Rb^{+/+}Pttg^{-/-}$ and $Rb^{+/-}Pttg^{-/-}$ mice. Each value represents mean $\pm$ SE ($n = 10-20$ mice/group). (**b**) Ki67 labeling index in 4-week-old $Rb^{+/+}Pttg^{+/+}$, $Rb^{+/-}Pttg^{+/+}$, $Rb^{+/+}Pttg^{-/-}$ and $Rb^{+/-}Pttg^{-/-}$ mice. Each value represents mean $\pm$SE (10 fields/animal; $n = 3$ animal/genotype analyzed); **, $P < 0.01$ in $Rb^{+/+}Pttg^{-/-}$ mice vs three other genotypes. (**c**) Development of pituitary tumors in $Rb^{+/+}Pttg^{+/+}$, $Rb^{+/-}Pttg^{+/+}$, $Rb^{+/+}Pttg^{-/-}$ and $Rb^{+/-}Pttg^{-/-}$ mice over time. $n =$ total number of animals sacrificed. Kaplan–Meier survival analysis (log-rank test) of the time of death with evidence of pituitary tumor in the different genotypes showed significant differences between $Rb^{+/-}Pttg^{-/-}$ and $Rb^{+/-}Pttg^{+/+}$ ($P < 0.01$), between $Rb^{+/-}Pttg^{-/-}$ and $Rb^{+/+}Pttg^{-/-}$ ($P < 0.05$), and between $Rb^{+/-}Pttg^{+/+}$ and $Rb^{+/+}Pttg^{-/-}$ mice ($P < 0.01$)

PTTG1 and Sp1, showing that the N-terminus binds to Sp1 whereas PTTG1 C-terminus is important for modulating Sp1 activity. Whether or not Sp1 binding on the cyclin D3 promoter is directly regulated by PTTG1 requires further investigation. PTTG1 and Sp1 also act in coordination in the G1/S phase transition. As Sp1 also regulates PTTG1 gene transcription, and the PTTG1 promoter has four potential Sp1 binding sites, the presence of an auto-feedback mechanism between Sp1 and PTTG1 has been suggested. Further support for this mechanism was gleaned from G1/S phase, genetically aberrant cells, including p21$^{-/-}$ HCT116 human colon cancer cells and $Rb^{+/-}$ MEFs, which confirmed the PTTG1 requirement for cell transformation and further suggested a G1/S phase role for PTTG1 in enabling cell transformation. Therefore, in addition to PTTG1 functioning as a securin protein playing a role in the G2/M phase transition, the protein plays a role in G1/S. PTTG1 regulation of the G1/S phase transition may therefore be an underlying mechanism for PTTG1 transforming activity.

The unique requirement of replicative endocrine tissue for *Pttg* implies that slow adaptive cell cycle progression is controlled by mechanisms distinct from the rapid cell cycle of skin or digestive tract regenerative tissues. This finding is consistent with clinical observations that tumors arising from the pituitary and from pancreatic islets rarely exhibit malignant phenotypes, whereas carcinomas frequently arise from actively regenerative tissues. Thus, in the pituitary of *Pttg*-null mice, activation of the ARF-p53-p21 senescence pathway may explain the observed pituitary hypoplasia as well as the underlying restraint of pituitary tumor development in *Pttg*-null mice.

Pttg-null mice are therefore reflective of a robust in vivo model for pituitary hypoplasia and premature senescence. Cross-breeding these animals with mice null for other cell cycle and senescence-related proteins, including p21 or p27, will provide further mechanistic insights into the pathogenesis of pituitary senescence. Thus, dissecting the respective roles of these proteins will allow a clearer definition of the mechanisms for trophic control of the pituitary gland cell cycle in health and disease. As senescence restrains cell growth, pituitary senescence may be protective for the exceedingly rarely encountered human malignant pituitary neoplasms. Pituitary senescence should therefore be considered as a trophic modulator underlying the overwhelmingly benign nature of pituitary tumors.

References

Abbud RA, Takumi I, Barker EM, Ren SG, Chen DY, Wawrowsky K, Melmed S (2005) Early multipotential pituitary focal hyperplasia in the alpha-subunit of glycoprotein hormone-driven pituitary tumor-transforming gene transgenic mice. Mol Endocrinol 19:1383–1391

Campisi J (2005) Suppressing cancer: the importance of being senescent. Science 309:886–887

Chesnokova V, Kovacs K, Castro AV, Zonis S, Melmed S (2005) Pituitary hypoplasia in Pttg-/- mice is protective for Rb+/- pituitary tumorigenesis. Mol Endocrinol 19:2371–2379

Heaney AP, Horwitz GA, Wang Z, Singson R, Melmed S (1999) Early involvement of estrogen-induced pituitary tumor transforming gene and fibroblast growth factor expression in prolactinoma pathogenesis. Nature Med 5:1317–1321

Melmed S (2003) Mechanisms for pituitary tumorigenesis: the plastic pituitary. J Clin Invest 112:1603–1618

Wang Z, Moro E, Kovacs K, Yu R, Melmed S (2003) Pituitary tumor transforming gene-null male mice exhibit impaired pancreatic beta cell proliferation and diabetes. Proc Natl Acad Sci USA 100:3428–3432

Yu R, Heaney AP, Lu W, Chen J, Melmed S (2000) Pituitary tumor transforming gene causes aneuploidy and p53-dependent and p53-independent apoptosis. J Biol Chem 275:36502–36505

Zou H, McGarry TJ, Bernal T, Kirschner MW (1999) Identification of a vertebrate sister-chromatid separation inhibitor involved in transformation and tumorigenesis. Science 285:418–422

p27 Regulation by Estrogen and Src Signaling in Human Breast Cancer

Joyce Slingerland

Summary

p27 is a key regulator of G1 to S phase progression. It inhibits cyclin E-Cdk2 in G0 and early G1 and promotes the assembly and activation of D-type cyclin-dependent kinases (cdks). While the p27 gene is rarely mutated in human cancers, p27 action is impaired in breast and other human cancers through accelerated p27 proteolysis, sequestration by cyclin D-cdks and p27 mislocalization in tumor cell cytoplasm. In cancers, reduced p27 protein reflects tumor de-differentiation and is associated with high histopathologic tumor grade. Reduced p27 is an indicator of poor patient outcome in primary human breast cancers. Following estrogen binding to the estrogen receptor (ER), p27 is regulated by the rapid transient ER-mediated activation of Src and signaling pathways. This review will focus on mechanisms of p27 regulation in normal cells and how Src activation may deregulate p27 in breast and other human cancers. The relevance of this pathway to antiestrogen therapy of breast cancers will also be reviewed.

Cell Cycle Regulation of G1 to S Phase

Progression through the cell cycle in mammalian cells is governed by a family of serine-threonine kinases, the cyclin-dependent kinases (cdks; Sherr 1994; Sherr and Roberts 1995; Reed et al. 1994). Cdk activity is regulated by positive and negative phosphorylation events, by binding of cyclins that serve as positive regulatory subunits and by negative regulators known as Cdk inhibitors (Reed et al. 1994; Sherr and Roberts 1999). G1- to S-phase progression is regulated by D-type cyclin cdks in early G1 and also by the activation of cyclin E- and cyclin A-dependent Cdk2 complexes promoting S phase entry and progression.

There are two families of cdk inhibitors in mammalian cells based on structure and cdk targets (Sherr and Roberts 1995; Slingerland and Pagano 2000; Sherr and Roberts 1999). Members of the inhibitor of cdk4 or INK4 family, $p15^{INK4B}$, $p16^{INK4A}$, $p18^{INK4C}$ and $p19^{INK4D}$, specifically bind cdk4 and cdk6, leading to cyclin D dissociation and kinase inhibition. The kinase inhibitor protein or KIP family members, $p21^{CIP1}$, $p27^{Kip1}$, and $p57^{Kip2}$, bind and inhibit cyclin E-Cdk2 and cyclin A-Cdk2 complexes. KIP members serve a dual function in G1 to S phase: they inhibit Cdk2 in G0 and early G1 but also facilitate the assembly and nuclear import of cyclin D-cdk4 and cdk6 complexes (LaBaer et al. 1997; Cheng et al. 1999).

[1] Braman Family Breast Cancer Institute, Sylvester Comprehensive Cancer Center, University of Miami School of Medicine, 1475 N.W. 12th Avenue (D8-4), Miami, FL 33136 USA

Melmed et al.
Hormonal Control of Cell Cycle
© Springer-Verlag Berlin Heidelberg 2008

Regulation of the Cell Cycle Inhibitor, p27[KIP1]

p27[Kip1] (hereafter p27) was first discovered as an inhibitor of cyclin E-Cdk2 and cyclin A-Cdk2 in cells arrested by transforming growth factor-β (TGF-β), by contact inhibition, and by lovastatin (Koff et al. 1993; Hengst et al. 1994; Polyak et al. 1994; Slingerland et al. 1994). p27 levels and activity increase in response to differentiation signals (Hengst and Reed 1996; Wang et al. 1996; Durand et al. 1997), to loss of adhesion to the extracellular matrix (Koyama et al. 1996; StCroix et al. 1998; Watanabe et al. 1996; Fang et al. 1996; Assoian 1997; Radeva et al. 1997), and on signaling by growth-inhibitory factors such as TGF-β (Koff et al. 1993; Polyak et al. 1994; Slingerland et al. 1994).

p27 acts in G0 and early G1 to inhibit cyclin E-Cdk2. p27 mRNA levels are constant throughout the cell cycle (Hengst and Reed 1996). p27 protein levels are maximal in quiescence, and estrogens and other mitogens stimulate p27 loss during G0- to S-phase progression, leading to cyclin E-Cdk2 activation. As cells exit quiescence, p27 that is bound to cyclin E must be degraded and newly synthesized cytoplasmic p27 together with p21 facilitate the assembly and import of cyclin D-cdk complexes. These effects are regulated by changes in the phosphorylation of p27 (Sheaff et al. 1997; Vlach et al. 1997; Ciarallo et al. 2002). On exit from quiescence, p27 levels fall rapidly due to a dramatic decrease in p27 translation and activation of p27 proteolysis.

Several mechanisms regulate p27 levels in G0 to S phase. p27 translation is maximal in G0 and falls abruptly as cells exit G0 (Gopfert et al. 2003; Hengst and Reed 1996; Agrawal et al. 1996). p27 proteolysis is activated early in G1, with the p27 t1/2 falling five- to eight-fold with passage from G0 to S phase (Pagano et al. 1995; Malek et al. 2001). At least four pathways regulate p27 degradation. In late G1 through S and into M phase cells, p27 is degraded by the SCF[Skp2] ubiquitin ligase. The recognition of p27 by the F-box protein Skp2, a component of the SCF-ubiquitin ligase, is activated by cyclin E-Cdk2-dependent p27 phosphorylation at T187 (Pagano et al. 1995; Sheaff et al. 1997; Vlach et al. 1997; Montagnoli et al. 1999). In early G1, T187-independent p27 proteolysis occurs by both Skp2-dependent (Malek et al. 2001) and Skp2-independent (Hara et al. 2001) pathways. In early G1, p27 phosphorylation at S10 promotes p27-CRM1 binding and nuclear p27 export (Rodier et al. 2001; Ishida et al. 2002; Connor et al. 2003). Proteolysis of cytoplasmic p27 in G1 involves KPC-1 ubiquitin ligase (Hara et al. 2005; Kamura et al. 2004). Rapid S10-independent p27 proteolysis may also occur in the nucleus in early G1 (Rodier et al. 2001). In quiescent cells, p27 proteolysis requires an intact p27-cyclin-Cdk binding motif (Besson et al. 2006).

The multiple mechanisms regulating p27 degradation act together to promote the loss of p27 that is required for S phase entry. The initial mitogen-stimulated, and potentially export-linked, phase of p27 degradation in early G1 would allow an incremental activation of cyclin E-Cdk2, which is then followed by rapid, progressive Cdk2 activation as activated cyclin E-Cdk2 mediates T187 phosphorylation-dependent p27 degradation in late G1 and S phase.

Switching p27 from Inhibitor to Substrate of Cyclin-Cdk2 Complexes

p27 is frequently inactivated in human cancers through accelerated p27 proteolysis. Reduced p27 is an independent marker of poor breast cancer prognosis that is correlated with higher tumor grade (reduced differentiation) and shorter disease-free survival (Alkarain et al. 2004). It has been recognized for at least a decade that p27 acts both as an inhibitor and as a substrate of cyclin E-Cdk2(Vlach et al. 1997; Sheaff et al. 1997). In G0 and early G1, p27 is tightly associated with cyclin E-Cdk2 and inhibits Cdk2 kinase activity. In contrast, in late G1, p27 becomes a substrate for cyclin E-Cdk2 and T187 phosphorylation of p27 promotes its binding to the SCF ligase mediating p27 degradation.

While cyclin E-Cdk2 phosphorylation of T187 targets p27 for SCF^{Skp2}-mediated proteolysis in late G1, the mechanisms permitting initial cyclin E-Cdk2 activation in early G1 despite the persistent abundance of its inhibitor presented a puzzle. We postulated that mitogen-mediated changes in p27 phosphorylation would lead to a shift in its function, weakening its association with Cdk2 to impair its inhibitor action on Cdk2. In cancers, activation of oncogenes could shift p27 phosphorylation such that a form of p27 with reduced Cdk2 inhibitory action could predominate. In cancer cell lines that overexpress ILK, Her2 or MEK oncogenes and in TGF-β resistant cells, we had observed altered p27 phospho-isoforms bound to cyclin-Cdk2 complexes that had reduced cyclin E-Cdk2 inhibitory function (Florenes et al. 1996; Ciarallo et al. 2002; Radeva et al. 1997; Donovan et al. 2001; Ciarallo et al. 2002). We postulated that such intermediate forms of p27 could facilitate the transition of p27 from inhibitor to substrate of cyclin E-Cdk2. Recent data from our group and from that of Hengst et al., now provide evidence that Abl and Src family-mediated tyrosine phosphorylation of p27 can reduce its inhibitory function for cyclin E-Cdk2 and facilitate p27 proteolysis (Chu et al. 2007; Grimmler et al. 2007).

Src is Frequently Activated in Human Breast Cancers

cSrc activates mitogenic signaling in normal and malignant cells (Thomas and Brugge 1997) to regulate cell proliferation, survival, metastasis, and angiogenesis (Ishizawar and Parsons 2004). Many human cancers, including breast, prostate, lung, colorectal, and ovarian carcinomas, show increased cSrc levels or activity when compared to adjacent normal tissues (Chen et al. 2006; Irby and Yeatman 2000; Biscardi et al. 2000). Src activation is often associated with increased disease stage (Frame 2002). We stained 482 primary human breast cancers for the phospho-activated form of Src (pY416-Src) and observed Src activation in 39% of these cancers (Chu et al. 2007).

Activating Src mutations are relatively rare in human cancers (Irby et al. 1999) and Src activation more commonly arises through oncogenic activation of receptor tyrosine kinases (RTK), including EGFR, Her2/ErbB2, IGF-1R, colony stimulating factor-1 and hepatocyte growth factor receptor. Reduced phosphorylation of its negative regulatory site, SrcY530, can also activate Src (Chen et al. 2006). The receptor tyrosine kinases, EGFR and Her2/ErbB2, both bind Src to catalyze mutual kinase activation. cSrc and EGFR family members are over-expressed in up to 70% of primary human breast cancers (Ishizawar and Parsons 2004; Biscardi et al. 2000). Inhibition of cSrc has been shown to block the effects of EGFR and Her2 on cell proliferation (Belsches-Jablonski et al. 2001; Biscardi et al. 1999). cSrc is also activated by estrogen binding to the estrogen receptor (ER) in human breast cancer cells. Estrogen:ER binding stimulates rapid transient recruitment of cSrc, Shc activation and MAPK signaling (Migliaccio et al. 1996). Liganded ER rapidly stimulates Src and further recruits receptor tyrosine kinases, Her2, EGFR (Chu et al. 2005) and IGF-1R (Song et al. 2004) to promote cell cycle progression.

Src Regulates the Inhibitory Action of p27 for Cdk2 and Facilitates p27 Proteolysis

We recently showed that cSrc can phosphorylate p27 in vitro and in vivo at both tyrosine 74 (Y74) and tyrosine 88 (Y88). p27 phosphorylation at the third tyrosine Y89 appears to be less important. A mutation converting p27Y89 to F reduced in vitro Src phosphorylation of p27 only modestly, and transfected p27Y89F showed little loss of overall phosphor-tyrosine content in vivo compared to p27WT. Hengst's group also demonstrated that p27 is phosphorylated by Abl kinase both in vivo and in vitro, primarily on Y88. Abl-phosphorylated p27 also did not show much reactivity with αpY89-specific antibodies (Grimmler et al. 2007). The Y89 site may modulate the efficiency of phosphorylation of the other two sites.

Src-phosphorylated p27 inhibited cyclin E-Cdk2 poorly in vitro, and Src transfection reduced cyclin E-Cdk2-bound p27. cSrc activation and Src-p27 co-precipitation preceded loss of p27-bound cyclin E-Cdk2 and loss of total p27 during G1 progression. Src induction increased pT187p27 and decreased p27 half-life.

Both p27Y74 and p27Y88 form contacts with Cdk2. The partial crystal structure of cyclin A-Cdk2-bound p27 showed that the region of p27 between Y74 and Y88 forms a clamp around the N-lobe of Cdk2 (Russo et al. 1996). Y74 is partially buried in the N-terminal β-sheet of Cdk2 and forms hydrophobic interactions with amino acids on the Cdk2 N-lobe. The aromatic ring of p27Y88 forms Van der Waals contacts with Cdk2, and its hydroxyl group forms hydrogen bonds with Cdk2 (Russo et al. 1996). p27Y88 occupies the catalytic cleft of Cdk2 and mimics the contacts made by the purine base of ATP (Russo et al. 1996).

Grimmler et al. (2007) provided NMR data to indicate that Y88 phosphorylation of p27 causes the 3_{10} helix of p27 to be ejected from the catalytic cleft of Cdk2. This finding provided an explanation for the reduced inhibitory action of p27 toward cyclin E-Cdk2 following Src kinase pre-treatment in vitro. Src-phosphorylated p27 showed a decreased steady state association with cyclin E-Cdk2 both in vitro and following co-transfection of p27WT with activated Src in vivo. Phosphorylation of Y74 together with Y88 appears to mediate loss of p27 inhibitory activity against cyclin E-Cdk2 and to promote dissociation of p27 from the Cdk2 complex. Taken together, these data support the notion that Src phosphorylation of p27 would impair the inhibitory action of p27 toward Cdk2, thereby facilitating cyclin E-Cdk2-dependent p27 proteolysis.

The specific intracellular tropisms of different tyrosine kinases toward the three pY sites on p27 remain to be determined. Activation of different tyrosine kinases may have different consequences in vivo on p27 function and stability. Further studies will be needed to define whether other cellular tyrosine kinases may also regulate these p27 sites in vivo in response to growth factor and hormonal stimulation. In certain cancers, constitutive Y88 phosphorylation, without concurrent Y74 phosphorylation, may permit activation of p27-associated cyclin-Cdk2. NMR structural data support this model (Grimmler et al. 2007).

p27 is Required for G1 Arrest by Antiestrogens in ER-positive Breast Cancers

As noted above, estrogen:ER binding has been shown to recruit Src and mediate activation of Ras, MEK and MAPK in a Shc-dependent manner (Migliaccio et al. 1996; Song et al. 2004). It has been known for over two decades that estrogen stimulates both G1 to S progression and recruitment of quiescent cells into cycle (Sutherland et al. 1983; Musgrove and Sutherland 1994). Estrogen stimulation of quiescent MCF-7 breast cancer cells causes a loss of p21 and p27 and of their binding to cyclin-Cdk2 to mediate cell cycle entry (Cariou et al. 2000).

Estrogen:ER-mediated Src activation appears to be a major effector of estrogen-driven 27 proteolysis. In both breast cancer lines and in 482 primary human breast cancers, we showed that Src activation was statistically correlated with reduced nuclear p27 levels. Src was rapidly activated by estrogen, and Src-p27 binding preceded the loss of cyclin E-Cdk2 from p27 and p27 proteolysis prior to S phase entrance. Src inhibition or transfection with Src siRNA increased p27 stability, and Src induction increased cellular pY74p27 and pT187p27 and reduced p27 stability. Src-mediated tyrosine phosphorylation of p27 would reduce the inhibitory action of p27 on cyclin E-Cdk2, liberating this kinase to phosphorylate its bound inhibitor at T187, promoting p27 ubiquitination and degradation (Pagano et al. 1995; Sheaff et al. 1997; Carrano et al. 1999). Hengst's group showed that tyrosine phosphorylation of p27 by Abl facilitated subsequent phosphorylation of p27 at T187 by cyclin A-Cdk2; thus pYp27 appears to be more readily phosphorylated by cyclin-Cdk2 compared to non-phosphorylated p27 (Grimmler et al. 2007). Thus, estrogen:ER activation of Src and tyrosine phosphorylation of p27 would drive the transition of p27 from inhibitor to substrate of cyclin-Cdk2 to promote p27 proteolysis.

p27 is Required for Therapeutic Effects of Antiestrogens

Approximately one third of human breast cancers express detectable levels of ERα protein and a majority of these can be treated effectively with ER-blocking drugs, including tamoxifen and fulvestrant, or aromatase inhibitors that deprive the tumor of circulating estrogen. Despite initial responsiveness, antiestrogen resistance usually develops and limits treatment efficacy. Estrogen deprivation or therapeutic ER blockade by the drugs tamoxifen or fulvestrant (ICI182780) leads to G0/G1 arrest with an accumulation of p21 and p27 and cyclin-Cdk2 inhibition. The cell cycle inhibitors, p27 and p21, are essential for the clinical efficacy of antiestrogens (Cariou et al. 2000; Carroll et al. 2000). When p21 or p27 expression was inhibited by antisense oligonucleotides, cells whose proliferation had been blocked by tamoxifen or ICI182780 re-entered the cell cycle. Depletion of either p21 or p27 mimicked the effect of estrogen to stimulate cell proliferation, indicating that both of these cell cycle inhibitors are essential mediators of the therapeutic cell cycle arrest by tamoxifen on breast cancer (Cariou et al. 2000).

Tamoxifen resistance is associated with high levels of expression or activation of the EGFR or Her2 family in primary breast cancers. EGFR and Her2/ErbB2 over-expression can confer tamoxifen resistance in culture (Benz et al. 1992; Pietras et al. 1995; Donovan et al. 2001). Src activation can also lead to tamoxifen resistance in breast cancer lines. Many hormone-resistant variants of MCF-7 have constitutive activation of the MEK/MAPK pathway (Donovan et al. 2001). In these lines, MEK inhibition with U0126 (Donovan et al. 2001) restored sensitivity to growth arrest by fulvestrant (ICI 182,780) and by tamoxifen. Antisense inhibition of p27 expression abolished this sensitivity, confirming the importance of p27 in this antiestrogen-mediated arrest. Moreover, transfection of constitutively activated MEK into hormone-sensitive cells induced loss of p27 and resistance to antiestrogen-mediated cell cycle arrest. Inhibition of Her2 and EGFR action using lapatinib has also been shown to restore response to tamoxifen or fulvestrant in three different partially resistant, ER-positive breast cancer cell lines (Chu et al. 2005).

A Role for Src Inhibitors in Treatment of Antiestrogen-resistant Breast Cancer

As noted above, antiestrogens that block the ER or deprive breast cancer cells of estrogen represent important therapeutic tools in breast cancer treatment. As noted above, p27 is required for G1 arrest by tamoxifen or estrogen deprivation (Cariou et al. 2000). Since activated receptor tyrosine kinases such as EGFR and Her2 recruit and activate Src, and given our observation that Src impairs the Cdk2 inhibitory action of p27, we tested whether Src inhibition could restore antiestrogen responsiveness to resistant lines.

In three different Src-activated, ER-positive breast cancer lines, Src-inhibitory doses of AZD0530, an orally available dual inhibitor of Abl and Src family kinases (Ple et al. 2004; Hiscox et al. 2006), or ER-saturating doses of tamoxifen, when given alone, each only partly reduced the percentage of S phase cells. AZD0530 together with tamoxifen increased p27, inhibition of cyclin E-Cdk2, and G1 arrest. In contrast, addition of AZD0530 together with tamoxifen to two other cell lines lacking Src activation did not reduce cell proliferation below that observed with tamoxifen alone. In our analysis of primary human breast cancers, we found that a little over one third of ER-positive cancers (127/339 or 39%) showed high Src activity scores (Chu et al. 2007). This proportion fits the known observation that about one third of ER-positive cancers will manifest de novo resistance to antiestrogen treatments. The observed rate of tamoxifen resistance and our observation that Src inhibition could restore cell cycle arrest by tamoxifen in partially resistant tumors raise the provocative possibility that Src inhibitors such as AZD0530 may have potential benefit in the treatment of human breast cancers. AZD0530 has shown good bioavailability and acceptable toxicity (AstraZeneca, unpublished data) and is currently in phase II trials as a single agent for metastatic human cancers. The potential for Src inhibitors to delay or reverse antiestrogen resistance or indeed to synergize therapeutically with antiestrogens in certain breast cancer contexts is under investigation. Src inhibitors may enhance the efficacy of antiestrogen therapy for women with this disease.

References

Agrawal D, Hauser P, McPherson F, Dong F, Garcia A, Pledger WJ (1996) Repression of p27kip1 synthesis by platelet-derived growth factor in BALB/c 3T3 cells. Mol Cell Biol 16:4327–4336

Alkarain A, Jordan R, Slingerland J (2004) p27 deregulation in breast cancer: prognostic significance and implications for therapy. J Mammary Gland Biol Neoplasia 9:67–80

Assoian RK (1997) Anchorage-dependent cell cycle progression. J Cell Biol 136:1–3

Belsches-Jablonski AP, Biscardi JS, Peavy DR, Tice DA, Romney DA, Parsons SJ (2001) Src family kinases and HER2 interactions in human breast cancer cell growth and survival. Oncogene 20:1465–1475

Benz CC, Scott GK, Sarup JC, Johnson RM, Tripathy D, Coronado E, Shepard HM, Osborne CK (1992) Estrogen-dependent, tamoxifen-resistant tumorigenic growth of MCF-7 cells transfected with HER2/neu. Breast Cancer ResTreat 24:85–95

Besson A, Gurian-West M, Chen X, Kelly-Spratt KS, Kemp CJ, Roberts JM (2006) A pathway in quiescent cells that controls p27Kip1 stability, subcellular localization, and tumor suppression. Genes Dev 20:47–64

Biscardi JS, Maa MC, Tice DA, Cox ME, Leu TH, Parsons SJ (1999) c-Src-mediated phosphorylation of the epidermal growth factor receptor on Tyr845 and Tyr1101 is associated with modulation of receptor function. J Biol Chem 274:8335–8343

Biscardi JS, Ishizawar RC, Silva CM, Parsons SJ (2000) Tyrosine kinase signalling in breast cancer: Epidermal growth factor receptor and c-Src interactions in breast cancer. Breast Cancer Res 2:203–210

Cariou S, Donovan JC, Flanagan WM, Milic A, Bhattacharya N, Slingerland JM (2000) Down-regulation of p21WAF1/CIP1 or p27Kip1 abrogates antiestrogen-mediated cell cycle arrest in human breast cancer cells. Proc Natl Acad Sci USA 97:9042–9046

Carrano AC, Eytan E, Hershko A, Pagano M (1999) SKP2 is required for ubiquitin-mediated degradation of the CDK inhibitor p27. Nature Cell Biol 1:193–199

Carroll JS, Prall OW, Musgrove EA, Sutherland RL (2000) A pure estrogen antagonist inhibits cyclin E-Cdk2 activity in MCF-7 breast cancer cells and induces accumulation of p130-E2F4 complexes characteristic of quiescence. J Biol Chem 275:38221–38229

Chen T, George JA, Taylor CC (2006) Src tyrosine kinase as a chemotherapeutic target: is there a clinical case? Anticancer Drugs 17:123–131

Cheng M, Olivier P, Diehl JA, Fero M, Roussel MF, Roberts JM, Sherr CJ (1999) The p21(Cip1) and p27(Kip1) CDK 'inhibitors' are essential activators of cyclin D-dependent kinases in murine fibroblasts. EMBO J 18:1571–1583

Chu I, Blackwell K, Chen S, Slingerland J (2005) The dual ErbB1/ErbB2 inhibitor, lapatinib (GW572016), cooperates with tamoxifen to inhibit both cell proliferation- and estrogen-dependent gene expression in antiestrogen-resistant breast cancer. Cancer Res 65:18–25

Chu I, Sun J, Arnaout A, Kahn H, Hanna W, Narod S, Sun P, Tan CK, Hengst L, Slingerland J (2007) p27 phosphorylation by Src regulates inhibition of cyclin E-Cdk2. Cell 128:281–294

Ciarallo S, Subramaniam V, Hung W, Lee JH, Kotchetkov R, Sandhu C, Milic A, Slingerland JM (2002) Altered p27(Kip1) phosphorylation, localization, and function in human epithelial cells resistant to transforming growth factor beta-mediated G(1) arrest. Mol Cell Biol 22:2993–3002

Connor MK, Kotchetkov R, Cariou S, Resch A, Lupetti R, Beniston RG, Melchior F, Hengst L, Slingerland JM (2003) CRM1/Ran-mediated nuclear export of p27(Kip1) involves a nuclear export signal and links p27 export and proteolysis. Mol Biol Cell 14:201–213

Donovan JC, Milic A, Slingerland JM (2001) Constitutive MEK/MAPK activation leads to p27(Kip1) deregulation and antiestrogen resistance in human breast cancer cells. J Biol Chem 276:40888–40895

Durand B, Gao FB, Raff M (1997) Accumulation of the cyclin-dependent kinase inhibitor p27/Kip1 and the timing of oligodendrocyte differentiation. EMBO J 16:306–317

Fang F, Orend G, Watanabe N, Hunter T, Ruoslahti E (1996) Dependence of cyclin E-CDK2 kinase activity on cell anchorage. Science 271:499–502

Florenes VA, Bhattacharya N, Bani MR, Ben-David Y, Kerbel RS, Slingerland JM (1996) TGF-beta mediated G1 arrest in a human melanoma cell line lacking p15INK4B: evidence for cooperation between p21Cip1/WAF1 and p27Kip1. Oncogene 13:2447–2457

Frame MC (2002) Src in cancer: deregulation and consequences for cell behaviour. Biochim Biophys Acta1602:114–130

Gopfert U, Kullmann M, Hengst L (2003) Cell cycle-dependent translation of p27 involves a responsive element in its 5′-UTR that overlaps with a uORF. Human Mol Genet 12:1767–1779

Grimmler M, Wang Y, Mund T, Cilensek Z, Keidel EM, Waddell MB, Jakel H, Kullmann M, Kriwacki RW, Hengst L (2007) Cdk-inhibitory activity and stability of p27Kip1 are directly regulated by oncogenic tyrosine kinases. Cell 128:269–280

Hara T, Kamura T, Nakayama K, Oshikawa K, Hatakeyama S, Nakayama KI (2001) Degradation of p27Kip1 at the G0-G1 transition mediated by a Skp2-independent ubiquitination pathway. J Biol Chem 276:48937–48943

Hara T, Kamura T, Kotoshiba S, Takahashi H, Fujiwara K, Onoyama I, Shirakawa M, Mizushima N, Nakayama KI (2005) Role of the UBL-UBA protein KPC2 in degradation of p27 at G1 phase of the cell cycle. Mol Cell Biol 25:9292–9303

Hengst L, Reed SI (1996) Translational control of p27Kip1 accumulation during the cell cycle. Science 271:1861–1864

Hengst L, Dulic V, Slingerland JM, Lees E, Reed SI (1994) A cell cycle-regulated inhibitor of cyclin-dependent kinases. Proc Natl Acad Sci USA 91:5291–5295

Hiscox S, Morgan L, Green TP, Barrow D, Gee J, Nicholson RI (2006) Elevated Src activity promotes cellular invasion and motility in tamoxifen resistant breast cancer cells. Breast Cancer Res Treat 97:263–247

Irby RB, Yeatman TJ (2000) Role of Src expression and activation in human cancer. Oncogene 19:5636–5642

Irby RB, Mao WG, Coppola D, Kang J, Loubeau JM, Trudeau W, Karl R, Fujita DJ, Jove R, Yeatman TJ (1999) Activating SRC mutation in a subset of advanced human colon cancers. Nature Genet 21:187–190

Ishida N, Hara T, Kamura T, Yoshida M, Nakayama K, Nakayama KI (2002) Phosphorylation of p27^{Kip1} on serine 10 is required for its binding to CRM1 and nuclear export. J Biol. Chem 277:14355–14358

Ishizawar R, Parsons SJ (2004) c-Src and cooperating partners in human cancer. Cancer Cell 6:209–214

Kamura T, Hara T, Matsumoto M, Ishida N, Okumura F, Hatakeyama S, Yoshida M, Nakayama K, Nakayama KI (2004) Cytoplasmic ubiquitin ligase KPC regulates proteolysis of p27(Kip1) at G1 phase. Nature Cell Biol 6:1229–1235

Koff A, Ohtsuki M, Polyak K, Roberts JM, Massague J (1993) Negative regulation of G1 in mammalian cells: inhibition of cyclin E-dependent kinase by TGF-beta. Science 260:536–539

Koyama H, Raines EW, Bornfeldt KE, Roberts JM, Ross R (1996) Fibrillar collagen inhibits arterial smooth muscle proliferation through regulation of Cdk2 inhibitors. Cell 87:1069–1078

LaBaer J, Garrett MD, Stevenson LF, Slingerland JM, Sandhu C, Chou HS, Fattaey A, Harlow E (1997) New functional activities for the p21 family of CDK inhibitors. Genes Dev 11:847–862

Malek NP, Sundberg H, McGrew S, Nakayama K, Kyriakidis TR, Roberts JM (2001) A mouse knock-in model exposes sequential proteolytic pathways that regulate p27Kip1 in G1 and S phase. Nature 413:323–327

Migliaccio A, DiDomenico M, Castona C, DeFalco A, Bontempo P, Nola E, Auricchio F (1996) Tyrosine kinas/p21ras/MAP-kinase pathway activation by estradiol receptor complex in MCF-7 cells. EMBO J 15:1292–300

Montagnoli A, Fiore F, Eytan E, Carrano AC, Draetta GF, Hershko A, Pagano M (1999) Ubiquitination of p27 is regulated by Cdk-dependent phosphorylation and trimeric complex formation. Genes Dev 13:1181–1189

Musgrove EA, Sutherland RL (1994) Cell cycle control by steroid hormones. Cancer Biol 5:381–389

Pagano M, Tam SW, Theodoras AM, Beer-Romero P, Del Sal G, Chau V, Yew PR, Draetta GF, Rolfe M (1995) Role of the ubiquitin-proteasome pathway in regulating abundance of the cyclin-dependent kinase inhibitor p27. Science 269:682–685

Ple PA, Green TP, Hennequin LF, Curwen J, Fennell M, Allen J, Lambert-Van Der Brempt C, Costello G (2004) Discovery of a new class of anilinoquinazoline inhibitors with high affinity and specificity for the tyrosine kinase domain of c-Src 1. J Med Chem 47:871–887

Pietras RJ, Arboleda J, Reese DM, Wongvipat N, Pegram MD, Ramos L, Gorman CM, Parker MG, Sliwkowski MX, Slamon DJ (1995) HER-2 tyrosine kinase pathway targets estrogen receptor and promotes hormone-independent growth in human breast cancer cells. Oncogene 10:2435–2446

Polyak K, Kato JY, Solomon MJ, Sherr CJ, Massague J, Roberts JM, Koff A (1994) p27Kip1, a cyclin-Cdk inhibitor, links transforming growth factor-beta and contact inhibition to cell cycle arrest. Genes Dev 8:9–22

Radeva G, Petrocelli T, Behrend E, Leung-Hagesteijn C, Filmus J, Slingerland J, Dedhar S (1997) Overexpression of the integrin-linked kinase promotes anchorage-independent cell cycle progression. J Biol Chem 272:13937–13944

Reed SI, Bailly E, Dulic V, Hengst L, Resnitzky D, Slingerland J (1994) G1 control in mammalian cells. J Cell Sci Suppl 18:69–73

Rodier G, Montagnoli A, Di Marcotullio L, Coulombe P, Draetta GF, Pagano M, Meloche S (2001) p27 cytoplasmic localization is regulated by phosphorylation on Ser10 and is not a prerequisite for its proteolysis. EMBO J 20:6672–6682

Russo AA, Jeffrey PD, Patten AK, Massague J, Pavletich NP (1996) Crystal structure of the p27Kip1 cyclin-dependent-kinase inhibitor bound to the cyclin A-Cdk2 complex. Nature 382:325–331

Sheaff RJ, Groudine M, Gordon M, Roberts JM, Clurman BE (1997). Cyclin E-CDK2 is a regulator of p27Kip1. Genes Dev 11:1464–1478

Sherr CJ (1994) G1 phase progression: cycling on cue. Cell 79:551–555

Sherr CJ, Roberts JM (1995) Inhibitors of mammalian G1 cyclin-dependent kinases. Genes Dev 9:1149–1163

Sherr CJ, Roberts JM (1999) CDK inhibitors: positive and negative regulators of G1-phase progression. Genes Dev 13:1501–1512

Slingerland J, Pagano M (2000) Regulation of the cdk inhibitor p27 and its deregulation in cancer. J Cell Physiol 183:10–17

Slingerland JM, Hengst L, Pan CH, Alexander D, Stampfer MR, Reed SI (1994) A novel inhibitor of cyclin-Cdk activity detected in transforming growth factor beta-arrested epithelial cells. Mol Cell Biol 14:3683–3694

Song RX, Barnes CJ, Zhang ZG, Bao YD, Kumar R, Santen RJ (2004) The role of Shc and insulin-like qrowth factor 1 receptor in mediating the translocation of estrogen receptor a to the plasma membrane. Proc Natl Acad Sci USA 101:2076–2081

St Croix B, Sheehan C, Rak J, Florenes VA, Slingerland J, Kerbel RS (1998) E-cadherin-dependent growth suppression is mediated by the cyclin-dependent kinase inhibitor p27^{Kip1}. J Cell Biol 142:557–571

Sutherland RL, Green MD, Hall RE, Reddel RR, Taylor IW (1983) Tamoxifen induces accumulation of MCF7 human mammary carcinoma cells in the G0/G1 phase of the cell cycle. Eur J Cancer Clin Oncol 19:615–621

Thomas SM, Brugge JS (1997) Cellular functions regulated by Src family kinases. Annu Rev Cell Dev Biol 13:513–609

Vlach J, Hennecke S, Amati B (1997) Phosphorylation-dependent degradation of the cyclin-dependent kinase inhibitor p27. EMBO J 16:5334–5344

Wang QM, Jones JB, Studzinski GP (1996) Cyclin-dependent kinase inhibitor p27 as a mediator of the G1-S phase block induced by 1,25-dihydroxyvitamin D3 in HL-60 cells. Cancer Res 56:264–267

Watanabe N, Hunter T, Ruoslahti E (1996) Dependence of cyclin E-cdk2 kinase activity on cell anchorage. Science 271:499–502

The Synergy of Two Ovarian Hormone-induced Enzymes in Human Mammary Carcinogenesis

Henri Rochefort[1]

Introduction

The mitogenic activity of estrogens mediated by the estrogen receptor α (ERα) is well established and explains most of the tumor promoter activity of estrogens in breast and endometrium. The basic understanding of the mechanism of trans-activation of ER by agonist and antagonist has led to the first therapies for solid tumors based on the inhibition of estrogen action with antiestrogens and estrogen production with aromatase inhibitors.

By contrast, for progesterone and synthetic progestins, discrepancies between results obtained in cell lines and in patients have slowed down the development of therapies based on the progesterone receptor pathway as a target.

I will review how studies over the last 30 years on the hormone-responsive human breast cancer cell lines have been translated to the clinic with prognostic and predictive markers in response to targeted therapies. A number of studies indicate several candidates among the hormone-responsive genes, some of them closely related to the control of cell cycle. I will concentrate on just two ovarian hormone-induced enzymes, associated with tumor growth and progression, that we have studied extensively, due to their abundance and their specificity of regulation by hormonal steroids. These proteins appear to be less directly and classically related to the initial control of cell cycle than the cyclin-dependant kinases, which are involved in the commitment of cells to enter an active cell cycle, leading to DNA synthesis and mitosis. However, they seem to be required for a tumor to grow and develop as a disease.

I will then summarize our more recent results, obtained in tissues directly collected from patients, concerning the variations in the level of these enzymes and the ovarian hormone receptors in the early steps of breast carcinogenesis.

Estrogens and Cathepsin D

An overview of studies from our laboratory on the hormone-responsive MCF7, ZR75 and T47D human breast cancer cell lines indicates that estrogens, via ERα, trigger a concerted phenotypic program that allows these cells to replicate and divide. Among numerous estrogen-induced proteins, intracellular transcription factors (fos, jun, c myc...) and cyclins (D and E) act as intracrine mitogens to trigger the entry of cells into an active G1 phase of the cycle (Fig. 1). As reported by several authors at this meeting, they directly affect the regulation of the cell cycle. Our laboratory

---

[1] CRLC Val d'Aurelle Bat B Cancer Research Center, 34298 Montpellier Cedex 5, France. *email:* henri.rochefort@valdorel.fnclcc.fr

Melmed et al.
Hormonal Control of Cell Cycle
© Springer-Verlag Berlin Heidelberg 2008

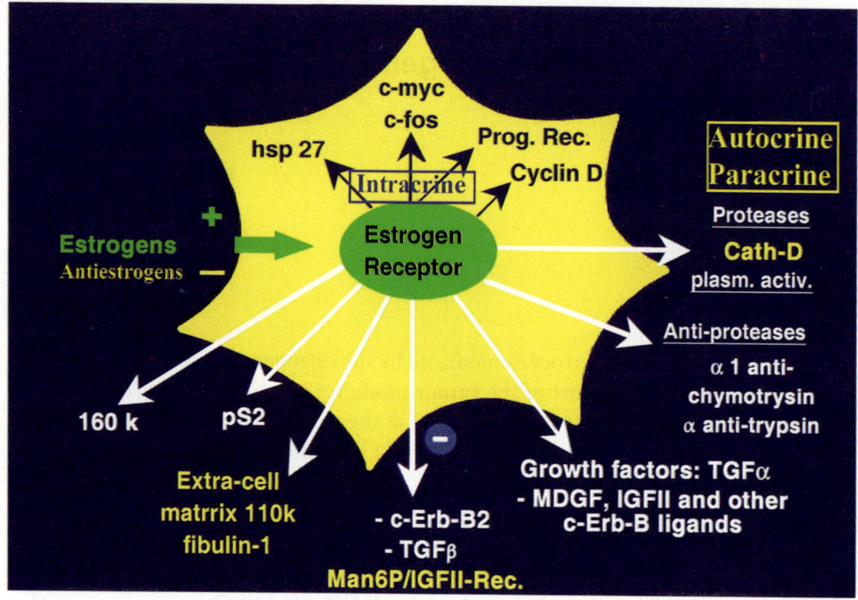

Fig. 1. In ERα-positive breast cancer cell lines, the levels of different mRNA and proteins have been shown to be stimulated by estrogens and inhibited by synthetic antiestrogens. Estradiol has also been shown to inhibit the synthesis of some proteins (*blue sign*). Several secreted proteins and peptides (*white arrows*) are potential paracrine or autocrine mitogens, such as growth factors and some proteases. The 160 kDa protein was not identified. Intra-nuclear proteins (*black arrows*) are potential intracrine factors. This is the case with cyclin D1, c-myc and c-fos, which are directly involved in the control of cell cycle. (Prog. Rec = progesterone receptor, plasm. activ = plasminogen activator)

has been more interested in estrogen-regulated secreted proteins, different of the classical growth factors, but having also the potential to act as autocrine and paracrine mitogens (Fig. 1). We have focussed on one prevailing, 52-kilodalton (kDa) secreted protein that is specifically induced by estrogens and inhibited by antiestrogens in parallel with the effect of these ER ligands on cell proliferation (Westley and Rochefort 1980: Chalbos et al. 1982; Vignon et al. 1986; Fig. 2). The 52 kDa protein was identified as the precursor of the lysosomal protease, Cathepsin D (cath D), after the quite unexpected discovery that its high level in primary breast cancer tissue extract was associated with an increased risk of clinical metastasis (Spyratos et al. 1989; Rochefort 1996). Cath D was not correlated with ERα in patients since it was also constitutively overexpressed in ER-negative cell lines (Rochefort et al. 1989). Cath D was also induced by growth factors, and its overexpression in human invasive breast cancers was found to be associated in several independent studies with an increased risk of relapse and metastasis (Rochefort et al. 2000; Ferrandina et al. 1997). This enzyme is a good potential therapeutic target because its overexpression stimulates the growth of cell lines and of experimental liver metastasis in vivo when rat embryo fibroblasts are stably transfected with cath D and then injected IV in nude mice (Garcia et al. 1990). Conversely, antisense cath D RNA stably transfected into MDA-MB231 human

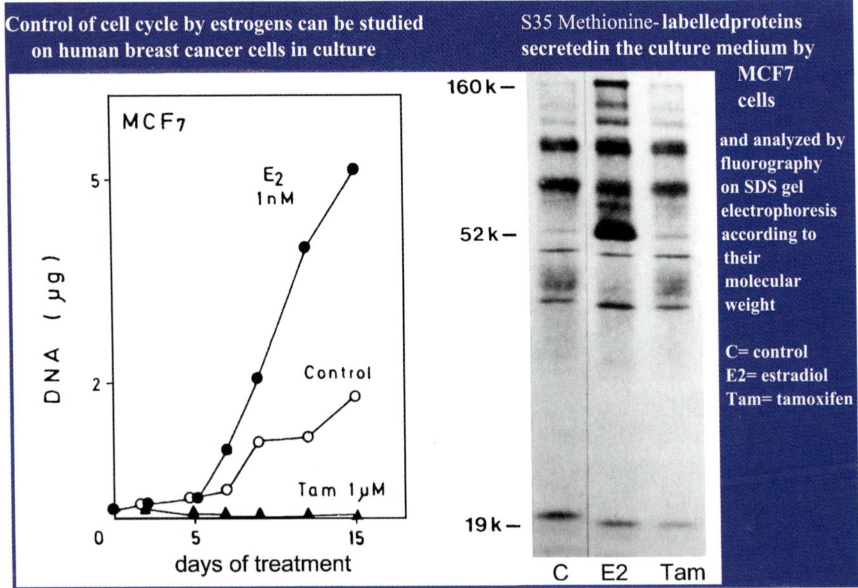

Fig. 2. How the correlation between the hormone regulation of cell proliferation and of the newly synthesized 52 kDa secreted protein led us to develop monoclonal antibodies and perform clinical studies in the hope of developing a circulating marker of hormone responsiveness in breast cancer. (Modified from Westley and Rochefort1980; Chalbos et al. 1982; Vignon et al. 1986)

breast cancer cells decreases both tumor growth in vitro and in vivo and experimental lung metastasis when injected in nude mice (Glondu et al. 2002). Several approaches can be used to inhibit its activity or production in experimental systems. However, a specific targeting of the inhibitors in cancer cells may be necessary since cath D is ubiquitously produced in several tissues. Its gene inactivation in KO mice induces apoptosis in thymus and necrosis in the highly proliferating cells of small intestine (Saftig et al. 1995). We described several potential mechanisms for this protein that, depending on the pH micro environment, can act as a protease digesting a number of potential substrates at an acidic pH or as a ligand activating a mitogenic pathway or displacing a mitogen to a receptor like the Man 6P/IGF2 receptor at a neutral pH (Rochefort and Liaudet-Coopman 1999; Rochefort et al. 2000). One interesting possibility is its action in phagolysosome where, by digesting engulfed extra cellular matrix, it could provide nutrients and space to the dividing cells, thus facilitating cancer cell survival and tumor growth (Montcourier et al. 1990, 1993, 1994). Digestion of phagocytosed extra cellular matrix was also demonstrated by the group of B Sloane by measuring in vivo the intracellular proteolytic activity of cathepsin B (Sameni et al. 2000).

In summary, for estrogens, the results of basic studies and clinical data have been consistent. Our understanding of the control of cell proliferation by estrogens and their antagonists has been translated to the clinic, leading to the routine use of targeted

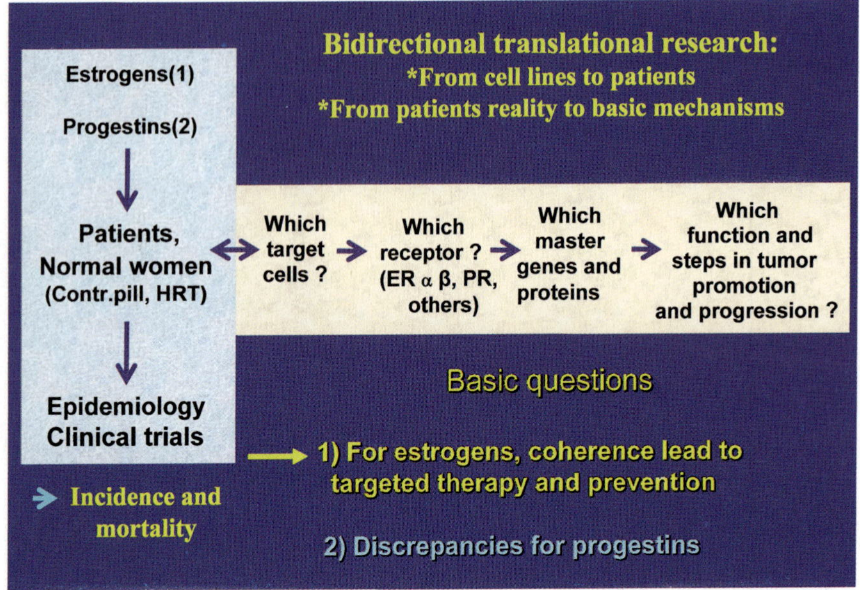

Fig. 3. The comparison between results obtained on monolayers of hormone responsive breast cancer cell lines and those obtained in vivo in patients showed coherence concerning the influence of estrogens and lead to predictive markers and targeted therapies. There were discrepancies with progestins, which are not yet fully understood

therapies with the antiestrogens or SERM and aromatase inhibitors. These drugs are also considered to be possible chemo-preventive agents for breast cancer (Fig. 3).

Progestins and Fatty Acid Synthase

By contrast, with progestins, many discrepancies have been observed between results obtained in cell lines and in patients. Several large-scale clinical studies, including randomized trials (Women's Health Initiative 2002, 2004) and cohort follow-up (Beral et al. 2003), have established that progestins, which protect endometrium against the mitogenic activity of estrogens, increase the risk of breast cancer in postmenopausal patients compared to estrogens alone. Moreover, the role of progesterone in breast development and tumorigenesis is well established in mice (Ismail et al. 2003). This finding is in contrast with studies in vitro on human cell lines, which indicate that progestins via their antiestrogen activity inhibit the estrogen-induced cell proliferation (Vignon et al. 1983). While the cause of the discrepancy between the in vitro and in vivo activities of progestins is not fully understood, Fatty Acid Synthase (FAS), an abundant progestin-induced enzyme, which we described and identified in these hormone-responsive cell lines (Chalbos et al. 1987a; Joyeux et al. 1990), provides a partial explanation for the tumor promoter activity of progestins. How progestins increase breast cancer risk is not clear, since via their antiestrogenic activity they inhibit rather than stimulate the growth of hormone-responsive cells in culture. However, the anti-progestin Ru 486 was anti-proliferative via the progesterone receptors (PR), (Bardon et al. 1985) and progestins also induce several proteins like the EGF receptor, the vascular endothelial growth factor (VEGF), etc. (Clarke and Sutherland 1990). One major difficulty is that many different progestins are being used in the treatment of menopause to prevent the deleterious effect of estrogens on the endometrium. Furthermore, some of the tumorigenic effect of progestins might be mediated by other receptors such as the androgen or estrogen receptors for the C19/C18 androstane progestins or the glucocorticoid receptor for the C21 pregnane progestins, like Medroxy Progesterone Acetate (MPA), that were mostly used in the USA in the WHI trials (Fig. 4).

Using hormone-responsive cell lines, we characterized several progestin-induced proteins. Some of them have been identified, like the secreted 18k and 43k proteins corresponding, respectively, to the GCDFP15 protein and the $Zn\alpha2$ glyco protein, which are both induced by the androgen receptor rather than PR because they are inhibited by anti-androgens and not by the anti-progestin RU486 (Chalbos et al. 1987a). Chalbos et al. (1987b) identified a very abundant cytoplasmic protein of 250 kDa as FAS, the multifunctional enzyme required for the synthesis of long chain fatty acids (Wakil 1989). We focused on this protein and prepared cDNA and specific antibody probes, allowing us to perform clinical studies. Comparing the different progestins, the weaker efficiency of progesterone compared to MPA or R5020 to induce FAS in cell lines can be explained by the higher metabolism of progesterone in these cells. This finding

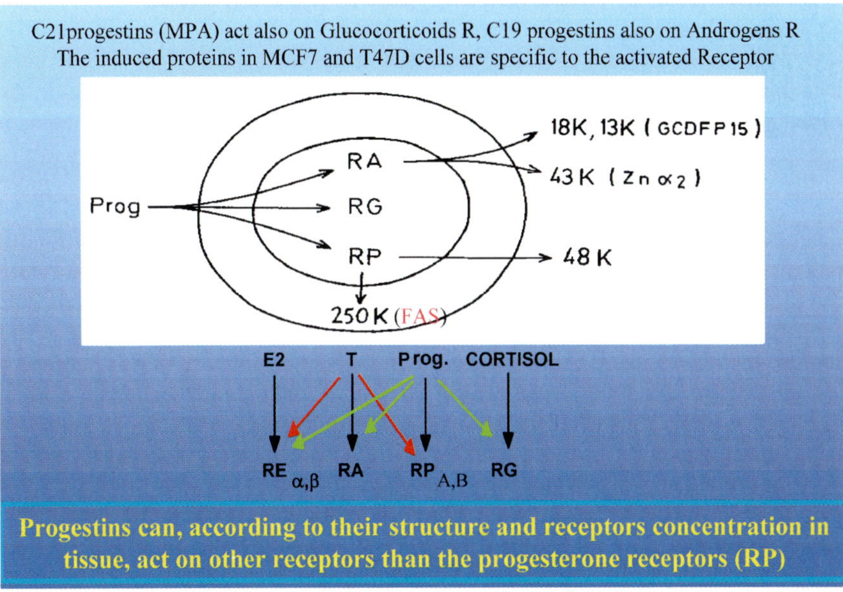

Fig. 4. Synthetic progestins according to their structure, can bind to 4 different classes of receptors as defined by their higher affinity for estrogens (RE), progesterone (RP), androgens (RA) or glucocorticoids (RG). This complexity may partly explain that the mechanism by which they increase breast cancer risk is not clear

might also explain the lower deleterious activity on breast of micronized progesterone compared to MPA in the French E3N–Epic/MGEN cohort (Fournier et al. 2005). Apparently, the same PR is responsible for inducing FAS, whatever the nature of the progestin is that activates this receptor. What is the significance of FAS overexpression in solid tumors? It is observed in both ER-positive and ER-negative invasive breast cancers, where it is associated with aggressiveness and increased risk of metastasis (Alo et al. 1996). FAS is required for the growth of human mammary tumor xenografts in the nude mice, as shown by the effect of FAS inhibitors and of si RNA (Kuhadja et al. 2000). It is likely that FAS provides endogeneous fatty acids to cancer cells that may not import them sufficiently from the circulation. Starting solid tumors and in situ carcinoma are often poorly vascularized. The endogenous synthesis of FAS would allow the synthesis of lipid membranes and supply energy needed for the continuously proliferating pre-cancer and cancer cells. While FAS is mostly produced normally in liver and in lactating mammary glands to provide fatty acids in the milk, its large overexpression in breast cancer cells, also observed in other solid tumors, appears to be crucial for cancer cell survival and growth in a nonfavorable micro environment. The inhibition of its expression by si RNA or of its activity by enzyme inhibitors decreases the growth of human mammary tumor xenografts in the nude mice. This enzyme is therefore considered to be a nutritional oncogene and a potential target for therapy (Kuhajda 2006).

Overexpression of These Two Enzymes in the Early Steps of Breast Carcinogenesis

Due to the genetic instability of invasive breast cancer, it is crucial to understand the initial mechanisms of sporadic breast carcinogenesis in an attempt to prevent them. It has become a health priority to stop the trend of the continuously increased incidence of this cancer in industrialized Western countries. Breast cancer incidence in France has increased two-fold in 20 years (Remontet 2003). There is no excellent in vitro model to study the mechanism of carcinogenesis of human sporadic breast cancer. In vitro human cell lines and animal models, even though useful, do not always reproduce the complexity of the processes taking place in women. Therefore, in the last decade, we have attempted to identify this mechanism in vivo by estimating directly in situ the varying expression of proteins involved in ovarian hormone action in mammary lesions at risk of developing breast cancer, compared with their expression in adjacent "normal" mammary glands (Dupont and Page 1985; Hartman et al. 2005). We asked two major questions concerning cath D and FAS:1) when are these enzymes overexpressed during the multisteps of carcinogenesis (Vogelstein B, Kinzler KW 1993) and 2) are they correlated in vivo with their corresponding steroid receptors, which also vary during carcinogenesis?

Several tissue banks have been used, beginning in 1994. The selection was based on the samples containing normal glands adjacent to the lesion. The cath D level was found to increase in apocrine cysts associated with proliferation and became quite significant in most high-grade ductal carcinoma in situ (DCIS) (Roger et al. 2000). However, there was a complete dissociation between the level of ERα and that of cath D, as previously reported in invasive breast cancer. These early studies led us to suggest a branched model of breast carcinogenesis, which is now supported and described further in several studies of transcriptomes using DNA micro arrays (Van't Veer et al. 2002), indicating that all breast cancers could be classified in at least five different clusters (Sorlie et al. 2003). Most of them (70–75%) are ERα-positive. However, the ERα-negative cancers are more aggressive and need other targets (Rochefort et al. 2003).

Interestingly, the number of ERβ positive cells decreased from normal glands to proliferative glands and in situ carcinoma. ERβ was expressed in both basal and luminal cells of normal duct, whereas ERα was only expressed in luminal cells (Roger et al. 2001; Fig. 5).

The overexpression of PR and FAS was dissociated because the FAS level increased as early as non-proliferating mastopathia and continued to increase to reach a maximum in both high- and low-grade DCIS (Esslimani-Sahla et al. 2006). Surprisingly, the PR level also increased early, before the increase in the ERα level, although the synthesis of this receptor is known to be normally up regulated by estrogens. While FAS was correlated with PR in early steps, both markers became independent and they were inversely correlated in ductal in situ carcinoma. This finding suggests that progestins via PR might increase FAS level in normal mammary glands, as previously reported (Joyeux

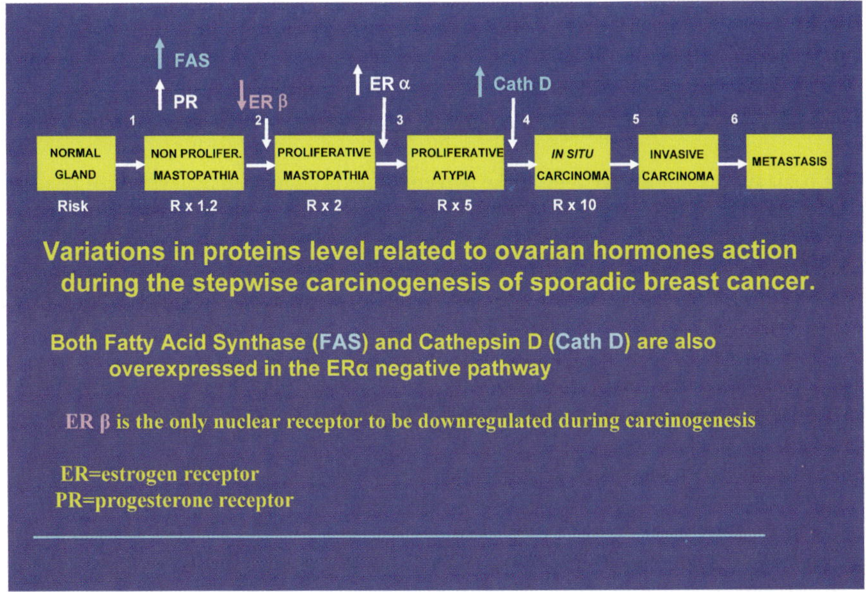

Fig. 5. A schematic and oversimplified representation of hormonal carcinogenesis of sporadic breast cancer showing (*1*) an early increase in PR and FAS, suggesting an increased responsiveness to progestins; (*2*) a continuous decrease in ERβ level, which is known to attenuate the mitogenic activity of the activated ERα by hetero-dimerization in breast cancer cells. This decrease associated with the increase in ERα level strongly suggests an increased sensitivity to estradiol as a mitogen; and (*3*) a dissociated expression of ERα and cath D in pre-invasive lesions, as shown previously in invasive carcinoma and in agreement with the up-regulation of this protease by other mitogens like growth factors

et al. 1990), but that this regulation is then bypassed by another type of regulation, possibly involving increased activities of MAP kinase and/or IP3 kinase, which are frequently detected in high-risk lesions (Yang et al. 2002). In this respect. FAS and HER2/Neu are highly correlated in in situ carcinoma. These results would suggest preventive action aiming to avoid systemic progestins in the hormone replacement therapy of menopause and eventually using FAS or progestin inhibitors in very high-risk lesions. Therefore, to summarize: 1) FAS and cath D levels are increased early in mammary carcinogenesis, as are their corresponding ovarian hormones receptors; 2) however, the Cath D level increased later than the FAS level and no correlation was found with ERα and with PR in high-grade DCIS. Other mechanisms appear to replace ovarian hormones to induce these genes in in situ carcinomas; 3) variations in ovarian steroid receptor levels and the inverse balance between ER α and ER β strongly suggest an early increased sensitivity of mammary glands to these hormones, even though other factors controlling receptor activity, like post-translational modifications, mutations, and variable dosage of co-modulators, could also play a role in the responsiveness to these hormones. These variations might explain why the tumoral tissue continues to proliferate even under low estrogen production after menopause. The mechanism of these variations, increased synthesis and/or decreased degradation, is unknown.

Discussion

How Might These Two Enzymes Synergize to Stimulate Tumor Growth?

The basis for a synergy between these two enzymes is suggested by a series of results:

1. FAS and cath D are induced, respectively, by estrogens (cath D) and progestins (FAS), with both hormones increasing, in synergy, breast cancer risk (WHI 2002 and 2004) and aggressiveness, as shown by their prognostic value as tissue markers;
2. FAS and cath D levels are increased successively at different steps of carcinogenesis and their levels are not correlated in in situ carcinomas;
3. their mechanism of action, even though not fully determined, clearly involves different and complementary pathways for the stimulation of tumor progression: cath D by providing amino acids, space and growth factor bio-availability and FAS by supplying the fatty acids, energy and lipid membrane needed to allow pre-cancer and cancer cells to proliferate;
4. the association of drugs targeted to them might therefore be considered in the future to treat breast cancer types that cannot respond to the biological drugs currently on the market, such as the SERMs and Herceptin.

What is the Relevance of These Overexpressed Enzymes to the Hormonal Control of the Cell Cycle, Which is the Topic of the Meeting?

It is clear that the commitment to enter the cell into DNA replication in G1 phase and cell division in G2 phase has been clarified via several cyclin-dependent kinases, as reported by others in this volume. However, ovarian hormones induce a concerted phenotypical program via a series of other proteins required for cancer cell autonomous growth, survival and division in unfavorable conditions due to the high pressure and poor vascularization in the tumor.

In this mini review, I have discussed only the two enzymes that my laboratory has extensively studied in the last 20 years, due to their abundance and the fact that they are specifically induced by one single type of steroid hormone, estrogen or progesterone.

There is considerable evidence that these enzymes are good potential targets for therapy in cancer subtypes that would relapse after the available adjuvant therapies.

Conclusions and Prospects

The cases of estrogen and progesterone appear to be quite different even though each type of steroid induces specifically the overexpression of a "housekeeping" enzyme involved in their synergistic role in tumor promotion.

With estrogens, we have reached a fairly good understanding of the concerted action, via several induced proteins and positive cross talk with growth factors, by which estrogens accelerate the cell cycle, first in facilitating the decision to enter cells into S phase and second in supplying amino acids, proteins and space that allow tumor cell growth, nutrition and survival.

Regarding progesterone, FAS should play a similar role by providing membrane lipids and energy. However, there are discrepancies between the results of clinical observations and cell line studies, suggesting an effect of progestins mediated by multicellular interactions. These discrepancies will require more work to understand in order to prevent these deleterious effects. These examples illustrate how the consistency of experimental studies in research laboratories and of clinical and histopathological observations in patients is crucial to achieve real progress in medicine.

Acknowledgements. I am indebted to many colleagues of the INSERM units Hormones and Cancers, who are quoted in the references. This review is dedicated to Françoise Vignon.

References

Alo PL, Visca P, Marci A, Mangoni A, Botti C, Di Tondo U (1996) Expression of fatty acid synthase (FAS) as a predictor of recurrence in stage I breast carcinoma patients. Cancer 77:474–482

Bardon S, Vignon F, Chalbos D, Rochefort H (1985) RU486, a progestin and glucocorticoid antagonist, inhibits the growth of breast cancer cells via the progesterone receptor. J Clin Endocrinol Metab 60:692–697

Beral V and the MWS group (2003) Breast cancer and hormone-replacement therapy in the Million Women Study. Lancet 362:419–427

Chalbos D, Vignon F, Keydar I, Rochefort H (1982) Estrogens stimulate cell proliferation and induce secretory proteins in a human breast cancer cell line (T47D). J Clin Endocrin Metab 55:276–283

Chalbos D, Chambon M, Ailhaud G, Rochefort H (1987a) Fatty acid synthetase and its mRNA are induced by progestins in breast cancer cells. J Biol Chem 262:9923–9926

Chalbos D, Haagensen D, Parish T, Rochefort H (1987b) Identification and androgen regulation of two proteins released by the T47D human breast cancer cells. Cancer Res 47:2782–2792

Clarke CL, Sutherland RL (1990) Progestin regulation of cellular proliferation. Endocr Rev 11:266–301

Dupont WD, Page DL (1985) Risk factors for breast cancer in women with proliferative breast disease. New Engl J Med 312:146–151

Esslimani-Sahla M, Thezenas S, Simony-Lafontaine J, Kramar A, Lavaill R, Chalbos D, Rochefort H (2007) Increased expression of Fatty Acid Synthase and Progesterone receptor in early steps of human mammary carcinogenesis. Int J Cancer 120(2):224–22

Ferrandina G, Scambia G, Bardelli F, Panici B, Mancuso S, Messori A (1997) Relationship between cathepsin-D content and disease-free survival in node-negative breast cancer patients: a meta-analysis. Br J Cancer 76:661–666

Fournier A, Berrino F, Riboli E, Avenel V, Clavel-Chapelon F (2005) Breast cancer risk in relation to different types of hormone replacement therapy in the E3N-EPIC cohort. Int J Cancer 114(3):448–454

Garcia D, Derocq D, Pujol P, Rochefort H (1990) Overexpression of transfected cathepsin D in transformed cells increases their malignant phenotype and metastatic potency. Oncogene 5:1809–1814

Glondu M, Liaudet-Coopman E, Derocq D, Platet N, Rochefort H, Garcia M (2002) Down-regulation of cathepsin-D expression by antisense gene transfer inhibits tumor growth and experimental lung metastasis of human breast cancer cells. Oncogene 21:5127–5134

Hartmann LC, Sellers TA, Frost MH, Lingle WL, Degnim AC, Ghosh K, Vierkant RA, Maloney SD, Pankratz VS, Hillman DW, Suman VJ, Johnson J, Blake C, Tlsty T, Vachon CM, Melton LJ 3rd, Visscher DW (2005) Benign breast disease and the risk of breast cancer. New Engl J Med 353:229–237

Ismail PM, Amato P, Soyal SM, DeMayo FJ, Conneely OM, O'Malley BW, Lydon JP (2003) Progesterone involvement in breast development and tumorigenesis – as revealed by progesterone receptor "knockout" and "knockin" mouse models. Steroids 68:779–787

Joyeux C, Chalbos D, Rochefort H (1990) Effects of progestins and menstrual cycle on fatty acid synthetase and progesterone receptor in human mammary glands. J Clin Endocrinol Metab 70:1438–1444

Kuhajda FP (2006) Fatty acid synthase and cancer: new application of an old pathway. Cancer Res 66:5977–5980

Kuhadja FP, Pizer ES, Li JN, Mani NS, Frehywot GL, Townsend CA (2000) Synthesis and antitumor activity of an inhibitor of fatty acid synthase. Proc Natl Acad Sci USA 97:3450–3454

Montcourrier P, Mangeat P, Salazar G, Morisset M, Sahuquet A, Rochefort H (1990) Cathepsin D in breast cancer cells can digest extracellular matrix in large acidic vesicles. Cancer Res 50:6045–6054

Montcourrier P, Valembois C, Rochefort H (1993) La présence de grandes vésicules acides est corrélée avec le pouvoir invasif in vitro de cellules mammaires cancéreuses. CR Acad Sci 316:421–424

Montcourrier P, Mangeat P, Valembois C, Salazar G, Sahuquet A, Duperray C, Rochefort H (1994) Characterization of very acidic phagosomes in breast cancer cells and their association with invasion. J Cell Sci 107:2381–2391

Remontet L, Esteve J, Bouvier AM, Grosclaude P, Launoy G, Menegoz F, Exbrayat C, Tretare B, Carli PM, Guizard AV, Troussard X, Bercelli P, Colonna M, Halna JM, Hedelin G, Mace-Lesec'h J, Peng J, Buemi A, Velten M, Jougla E, Arveux P, Le Bodic L, Michel E, Sauvage M, Schvartz C, Faivre J (2003) Cancer incidence and mortality in France over the period 1978–2000. Rev Epidemiol Santé Publ 51:3–30

Rochefort H (1996) The prognostic value of Cathepsin-D in Breast Cancer. A long road to the clinic. Eur J Cancer 32A:7–8

Rochefort H, Liaudet-Coopman E (1999) Cathepsin D in cancer metastasis: A protease and a ligand. APMIS 107:86–95

Rochefort H, Cavaillès V, Augereau P, Capony F, Maudelonde T, Touitou I, Garcia M (1989) Overexpression and hormonal regulation of pro-cathepsin D in mammary and endometrial cancer. J Steroid Biochem 34:77–182

Rochefort H, Garcia M, Glondu M, Laurent V, Liaudet E, Rey JM, Roger P (2000) Cathepsin D in breast cancer: mechanisms and clinical applications, a 1999 overview. Clin Chim Acta 291:157–170

Rochefort H, Glondu M, Sahla ME, Platet N, Garcia M (2003) How to target estrogen receptor-negative breast cancer? Endocrine Rel Cancer 10:261–266

Roger P, Daures JP, Maudelonde T, Pignodel C, Gleizes M, Chapelle J, Marty-Double C, Baldet P, Mares P, Laffargue F, Rochefort H (2000) Dissociated overexpression of cathepsin-D and estrogen receptor alpha in preinvasive mammary tumors. Human Pathol 31:593–600

Roger P, Esslimani Sahla M, Makela S, Gustafsson J-Å, Baldet P, Rochefort H (2001) Decreased expression of estrogen receptor β protein in proliferative preinvasive mammary tumors. Cancer Res 61:2537–2541

Saftig P, Hetman M, Schmahl W, Weber K, Heine L, Mossmann H, Köster A, Hess B, Evers M, Von Figura K, Peters C (1995) Mice deficient for the lysosomal proteinase cathepsin D exhibit progressive atrophy of the mucosa and profound destruction of lymphoid cells. EMBO J 14:3599–3608

Sameni M, Moin K, Sloane BF (2000) Imaging proteolysis by living human breast cancer cells. Neoplasia 2:495–504

Sorlie T, Tibshirani R, Parker J, Hastie T, Marron JS, Nobel A, Deng S, Johnsen H, Pesich R, Geisler S, Demeter J, Perou C, Lønning PE, Brown PO, Børresen-Dale AL, Botsteinet D (2003) Repeated observation of breast tumor subtypes in independent gene expression data sets. Proc Natl Acad Sci USA 100:8418–842

Spyratos F, Brouillet JP, Defrenne A, Hacene K, Rouesse J, Maudelonde T, Brunet M, Andrieu C, Desplaces A, Rochefort H (1989) Cathepsin-D: an independant prognostic factor for metastasis of breast cancer. Lancet 8672:1115–1118

van 't Veer LJ, Dai H, van de Vijver MJ, He YD, Hart AA, Mao M, Peterse HL, van der Kooy K, Marton MJ, Witteveen AT, Schreiber GJ, Kerkhoven RM, Roberts C, Linsley PS, Bernards R, Friend SH (2002) Gene expression profiling predicts clinical outcome of breast cancer. Nature 415:530–536

Vignon F, Bardon S, Chalbos D, Rochefort H (1983) Antiestrogenic effect of R5020, a synthetic progestin in human breast cancer cells in culture. J Clin Endocrinol Metab 56:1124–1130

Vignon F, Capony F, Chambon M, Freiss G, Garcia M, Rochefort H (1986) Autocrine growth stimulation of the MCF7 breast cancer cells by the estrogen-regulated 52k protein. Endocrinology 118:1537–1545

Vogelstein B, Kinzler KW (1993)The multistep nature of cancer. Trends Genet 9:138–14

Wakil S (1989) Fatty acid synthase, a proficient multifunctional enzyme. Biochemistry 28:4523–4530

Westley BR, Rochefort H (1980) A secreted glycoprotein induced by estrogen in human breast cancer cell lines. Cell 20:353–362

Writing Group for the Women's Health Initiative Investigators (2002) Risks and benefits of estrogen plus progestin in healthy postmenopausal women. JAMA 288:321–333

Writing Group for the Women's Health Initiative Investigators (2004) Effects of conjugated equine estrogen in postmenopausal women with hysterectomy. JAMA 291:1701–171

Yang YA, Han WF, Morin PJ, Chrest FJ, Pizer ES (2002) Activation of Fatty acid synthesis during neoplastic transformation: role of mitogen-activated protein kinase and phosphatidylinositol 3-kinase. Exp Cell Res 279:80–90

Steroid Receptors, Stem Cells and Proliferation in the Human Breast

Hannah Harrison[1], *Rebecca Lamb*[1], and *Robert B. Clarke*[1]

Abstract

This review summarises the current evidence for adult tissue stem cells in the human breast and examines the role of stem cells, steroids and self-renewal signalling pathways in normal human breast development.

The development of the mammary gland is a complex and lengthy process, beginning during embryogenesis and not reaching full functionality until pregnancy and lactation. The gland goes through massive apoptosis and remodelling at the cessation of weaning to resemble once again the non-pregnant gland. The staged and cyclic development of the mammary gland is controlled by ovarian steroids, and it is thought to only be possible due to the presence of adult stem cells. Adult stem cells are self-renewing cells that give rise to all of the functional, differentiated cells of the tissue/organ. The process of self-renewal and stem cell maintenance is tightly controlled and thought to be regulated by signalling, such as through the Notch receptor pathway.

Long-lived self-renewing stem cells are thought to be the targets of the accumulated mutations that lead to cancer. To fully determine the role of these cells and their signalling in cancer, it is important to first elucidate their function in normal breast development. A great deal of research has now been carried out into the identification and isolation of normal mammary stem cells.

Mammary Gland Development

Development begins during embryogenesis, with the formation of a rudimentary ductal system. The gland remains virtually unaltered throughout childhood, with no further development occurring until the hormonal changes of puberty (Naccarato et al. 2000). During this period of development, the ductal branches that are formed during embryogenesis grow and divide to form branching ductal bundles with terminal end buds (TEB; Smith and Neville 1984). The TEBs are a major site of proliferation, and at menarche the terminal duct lobuloalveolar units (TDLU) develop from this site but remain in a resting state until the onset of pregnancy and lactation (Vogel et al. 1981).

[1] Breast Biology Group, Division of Cancer Studies, Faculty of Medicine and Human Sciences, University of Manchester, Paterson Institute for Cancer Research, Wilmslow Road, M20 4BX, Manchester, UK

email: robert.clarke@manchester.ac.uk

Melmed et al.
Hormonal Control of Cell Cycle
© Springer-Verlag Berlin Heidelberg 2008

Further development occurs during pregnancy, with accelerated development of the TDLUs as the numbers of epithelial cells and alveoli within the lobules increase in preparation for lactation (Hovey et al. 2002). It is at the point of lactation that the breast can be said to reach full development and function. Following lactation, there is massive apoptosis and remodelling of the tissue that will then resemble, once again, the gland in its non-pregnant state (Allan et al. 2004; Furth et al. 1997).

The complex branching structure of lobules is lined by two distinct layers of cells: luminal epithelial cells, which synthesise milk, and myoepithelial cells, which form the basal layer of the ducts (Daniel and Smith 1999). There is strong evidence that these two different cell types arise from a common, pluripotent stem cell.

Furthermore, the ability of the mammary gland to then pass through multiple cycles of proliferation and apoptosis during pregnancy, lactation and involution is thought to be made possible by stem cells residing in the tissue.

Steroid Receptor Expression

The ovarian steroids, oestrogen and progesterone, are known to play a vital role in the staged development of the mammary gland, acting through specific nuclear receptors on target cells. These cells, which represent less than 20% of the epithelium, express oestrogen receptor alpha (ERα) and progesterone receptor (PR), invariably together, and are known to be located in the luminal epithelia of the ductal and lobular structures (Peterson et al. 1987). ERα/PR-expressing cells are non-proliferative and have been shown to co-express transforming growth factor beta (TGFβ), which may act as an autocrine signal to stop proliferation (Ewan et al. 2005). As ERα/PR-negative cells, which are often adjacent to receptor-positive cells (Brisken et al. 1998; Clarke et al. 1997), cannot respond directly to hormonal signalling, it is thought that their proliferation is controlled by a paracrine mediator of the systemic signal.

A recent study showed that in an ERα homozygous knockout mouse model (ERα$^{-/-}$), no mammary development occurred beyond the formation of the rudimentary structure seen at embryogenesis. If ERα$^{-/-}$ cells are mixed with wild type ERα cells before they are engrafted into a recipient cleared fat pad, however, the ERα$^{-/-}$ cells are able to proliferate and contribute to normal mammary gland development (Mallepell et al. 2006). These findings supported the theory that a paracrine signal is released from the non-dividing ERα/PR-positive cells in response to the steroidal signal that causes proliferation of the receptor-negative cells.

In 2006, Wilson et al. (2006) examined the effect of oestrogen on gene expression in mouse models. A number of genes were differentially expressed, including Amphiregulin (AR), a ligand of the epidermal growth factor (EGFR) signalling pathway. This finding supports the long-held theory that the EGFR signalling pathway plays a role in this paracrine signalling (Woodward et al. 1998). This belief was founded on the evidence that, when delivered as pellets, EGFR ligands, such as epidermal growth factor (EGF), transforming growth factor alpha (TGFα) and AR, were able to induce cell proliferation in ovariectomied mice (Coleman et al. 1988) in the same way as oestrogen (Daniel et al. 1987). More recently, the roles of EGF, TGFα and AR were shown in mammary development within mammopoeisis, lactogenesis and ductal outgrowth effected in knockout mice (Luetteke et al. 1999).

Recent studies have supported this theory but have shown specifically that AR is the paracrine mediator of ERα signalling (Brisken et al. 1998). Upon injection of oestrogen into the mammary glands of ovariectomied mice, AR expression was increased approximately 50-fold whereas other EGFR ligands were not affected. To test this theory, further mouse models were developed that completely lack AR. In these mice, development was impaired and proliferation at puberty was stopped.

Adult Tissue Stem Cells

Adult tissue stem cells (ASCs) are unspecialised and are defined by their ability to both self-renew throughout the life span of the host/organ and to give rise to all of the specialised cells types (Mayhall et al. 2004). Self-renewal is a process by which a stem cell (SC) divides, maintaining the SC population either asymmetrically or symmetrically (Morrison 1997). Asymmetric division is the production of one new SC and one, more differentiated, daughter cell. This daughter cell will then go on to generate committed cells that will undergo terminal differentiation down a specific cell lineage. Symmetric division occurs where either both daughter cells remain as undifferentiated SCs or both differentiate and the SCs are lost. The process of self-renewal is tightly controlled so that the expansion of SCs during normal tissue homeostasis is restricted through asymmetric division, but in times of need, such as tissue development, replacement or repair, symmetric division will occur to replenish the SC compartment (Potten and Loeffler 1990).

The first evidence of ASCs came in the 1960s with the discovery of haematopoietic stem cells (HSCs) in bone marrow (Siminovitch et al. 1963). HSCs were found to be multipotent, having the ability of multi-lineage differentiation generating precursor cells that can differentiate into all mature blood cells. This ability was shown using retroviral tagging of individual bone marrow cells before transplantation into a lethally irradiated mouse (Bonnet 2003). The retrovirus was genetically modified, confining it to the originally tagged cell and its progeny and allowing easy identification of the cells produced from the tagged cell. Using this method, it was possible to show that all classes of blood cell derive from a common cell (Bonnet 2003; Jordan and Lemischka 1990).

Since then ASCs have been found in a number of tissues and, although HSCs are the most well characterised to date, a great deal of progress has been made in their characterisation.

Mammary Epithelial SCs

In 1959, studies in which tissue was removed from the mammary gland of a donor mouse and transplanted into the cleared mammary fat pad of a recipient suggested the presence of SCs in the tissue (DeOme et al. 1959). This tissue section was seen to regenerate the tissue of origin completely, developing an entire and functioning mammary tree. This finding was supported when, in 1971, Daniel and Young (Daniel and Young 1971) showed that tissue sections could reconstitute the gland in up to eight serial transplantations. The authors concluded that there were indeed SCs present but the eventual growth senescence of serially transplanted epithelium was due to the number of divisions of these rare cells that had taken place. More recently, clonal dominant populations within the tissue were shown to be responsible for the outgrowth using mammary epithelia that was marked with mouse mammary tumour virus (MMTV; Kordon and Smith 1998). This work also showed that serial transplantation of these cells continually recapitulated the gland, demonstrating the self-renewing and multi-potent characteristics of the cells.

Strong evidence for the existence of mammary ASC came in two studies published in 2006 in which cells were sorted on their cellular characteristics (Shackleton et al. 2006; Stingl et al. 2006). One study reported that a single mouse mammary cell taken from a subpopulation that was negative for known lineage markers (Lin$^-$) and positive for the cell surface markers CD29 and CD24 (Lin$^-$/CD29hi/CD24$^+$) was able to recon-stitute the cleared mammary fat pad of a recipient mouse. One cell in 64 from this subpopulation had the ability to produce the normal, heterogeneous cell make-up of the gland and morphologically distinguishable structures like ducts and alveolar. An-other study showed that this subpopulation could be further enriched using a CD49f+ sort with one in 20 mouse mammary cells from this population, having the ability to regenerate the entire gland and showed evidence of up to 10 symmetrical self-divisions.

To further study the phenotypic and functional properties of these putative ASC, it is necessary to both identify and isolate them. This process has proven technically difficult due to the small number of ASCs in the tissue and the lack of SC-specific markers.

Side Population Analysis

The ability of some cells to efflux fluorescent dyes, such as Hoechst 33342, through membrane transporter proteins allows flow cytometric analysis and isolation of these cells that are known as a side population (SP). SP cells in the haematopoietic system represent the stem cell compartment, as they can reconstitute the bone marrow of lethally irradiated mice (Goodell et al. 1997).

This method has been used to study the SP of mouse mammary epithelium, and these cells, which make up less than 3% of the cell total, are enriched for putative SC markers such as Sca1 and α6-integrin (Welm et al. 2002; Liu et al. 2004). A similar SP has been identified in human breast tissue with SP proportions of up to 5% of the total epithelium (Alvi et al. 2003; Clarke et al. 2005; Contu et al. 2003). SP cells have been shown to be multipotent by their ability to differentiate into both luminal and myoepithelial cells in 3D matrigel culture (Clarke et al. 2005) and their ability to regenerate the gland on transplantation (Alvi et al. 2003). This finding suggests that human breast SP cells are also SC-like.

Once SP cells were isolated, it was possible to study markers of differentiation and to identify putative SC markers in the SP of human breast (Clarke et al. 2005; Clayton et al. 2004). It was shown that 70% of the SP cells lacked MUC1 and CALLA/CD10, markers of luminal and myoepithelial differentiation, whereas the majority of non-SP cells showed expression of one of these markers. A distinct, intermediate cell type has been shown previously and was suggested to be the common precursor to the luminal and myoepithelial cell lineages (Smith and Neville 1984; Chepko and Smith 1977). This intermediate cell was distinguished by pale cytoplasmic staining and reduced organelle number under light microscopy. The infrequent occurrence of these cells and their undifferentiated characteristics made them the focus of studies searching for the breast SC. Evidence supporting the theory that these cells are SC-like can be seen in a number of studies. When mammary epithelial cells are extracted from nulliparous mice and placed in culture in vitro, it is the pale cells that begin dividing first (Smith and Medina 1988). It has also been shown that pale staining cells are depleted in serially transplanted mammary epithelium and that their disappearance coincides with growth cessation (Smith et al. 2002), as previously reported by Daniel and Young 1971.

It was also shown from SP analysis that the expression of ERα was increased six-fold in the SP cells compared to the non-SP (Clarke et al. 2005). In this study, a number of putative SC markers were identified by QRT-PCR, for example p21, which is known to maintain cellular quiescence, and Musahi-1 (Msi1), an RNA binding protein thought to play a role in asymmetric division. These genes were reported to have two-fold and six-fold increases in expression within the SP, respectively. Interestingly, these markers were also shown to co-localise with ERα in mammary epithelial cells by dual label immunofluorescence. They were not, however, seen to co-localise together, suggesting two subsets of ERα-positive SP cells. Furthermore, the proliferation marker Ki67 was not seen to be present in the SP cells, supporting the theories that ERα-positive cells do not proliferate and that the tissue-specific SC is quiescent (Clarke et al. 2005).

Steroid Receptor Expression and SCs

SP analysis seems to suggest that stem-like cells are steroid receptor-positive and undifferentiated as these cells have been shown to be multipotent and to form mammospheres in culture. In contrast to this finding, however, studies using single cell transplantation for mouse mammary reconstitution showed the SC-enriched sub-population of Lin$^-$/CD24$^+$/CD49f+ cells are ERα-negative (Asselin-Labat et al. 2006).

This contradiction can be explained by studies of the cell types seen at different developmental stages in the human breast. Studies of early foetal breast tissue show that there are no ERα-positive cells present (Keeling et al. 2000) but, after 30 weeks of gestation, ERα-positive epithelial cells can be identified and are present in high numbers shortly after birth.

Work has been carried out to further define the steroid receptor status of mammary SCs using pulse chase experiments of DNA label-retaining cells (LRC) in mice. LRCs are cells that, following labelling, have divided only a small number of times and can, therefore, be assumed to be stem-like. Different results have come from these studies, depending on the developmental timing of the labelling. When labelling was carried out during puberty, the majority of LRCs were reported to be steroid receptor-negative (Welm et al. 2002) whereas when labelling occurred post-puberty, 95% of LRCs were steroid receptor-positive (Zeps et al. 1998).

These findings seem to suggest that different cell lineages in the normal breast are derived from SCs with different potentials at different developmental time points, with the ERα-negative cells being the most primitive SCs. Figure 1 shows a schematic view of the breast epithelial cell lineages.

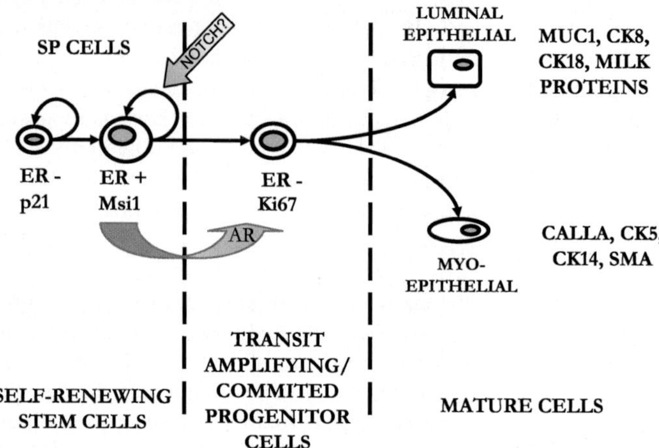

Fig. 1. A putative model of cellular hierarchy in the normal human breast. Side population (SP) cells are identified by their ability to efflux Hoechst dye and have been identified as both steroid receptor-negative and -positive cells. It has been suggested that the more primitive developmental stem cell is ERα negative. The label-retaining cells (LRC) and SP studies suggest that some of the stem cells are ERα positive and these may be the short-term adult stem cells. Quiescence is maintained due to the expression of p21, and the involvement of Msi1 with Delta/Notch signalling may allow self-renewal by asymmetric division. After a number of divisions, the transit amplifying/committed progenitor cells will leave the cell cycle and differentiate into either luminal or myoepithelial cell lineages. The differentiated myo- and luminal epithelial cells are characterised by proteins such as common acute lymphoblastic leukemia (CALLA) or mucin 1 (MUC1), respectively. **ER**, oestrogen receptor; **CK**, Cytokeratin; **Ki67**, marker of proliferation; **SMA**, smooth muscle actin; **AR**, amphiregulin

The Role of Notch Signalling

Notch was initially identified in Drosophila almost a century ago, when a mutation in the gene was seen to result in a mutant fly with notches in its wings (Morgan 1917). Figure 2 shows the Notch signalling pathway, which has been shown to act as a mediator of cell–cell communication in many tissues (Stylianou et al. 2006; Lai 2004) and is thought to play a role in SC self-renewal as well as cell fate, apoptosis, proliferation and migration (Politi et al. 2004). Notch has been well documented in cell fate decisions in both Drosophila and vertebrate embryogenesis (Kopan 2002) and has been described as a neurogenic gene (Lai 2004) due to its effect in the early stages of neural development.

Notch has been characterised as a large protein (Lai 2004) of which there are four different mammalian homologues: Notch 1–4. The proteins are made up of a ligand binding extracellular domain that is non-covalently bound to an intracellular domain (Callahan and Egan 2004). The Notch extracellular domain (NECD) acts as a receptor to five known ligands of the DSL (Delta, Serrate, LAG) superfamily, which include Delta-like 1, 3 and 4 (DLLA 1, 3, 4) and Jagged 1 and 2 (JAG 1, 2; Mumm et al. 2000). These ligands must be bound to a neighbouring cell membrane to activate the Notch pathway, and secreted ligands have been shown to act as blockers of ligand-receptor interactions

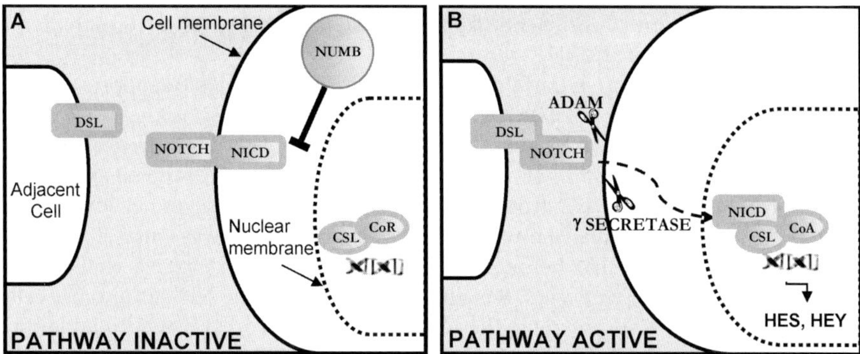

Fig. 2. A schematic overview of the Notch signalling pathway. (**A**) When the Notch receptor is unbound or Numb is active, the signalling pathway is inactive/inhibited. The Notch intracellular domain (NICD) remains bound to the cell membrane. Within the nucleus, the CBf1, suppressor of hairless, LAG2 (CSL) protein complex remains bound to a co-repressor. (**B**) Notch signalling is activated when the Notch receptor binds to a Delta, Serrate, Lag-2 (DSL) ligand on the surface of a neighbouring cell. Upon binding, the extracellular domain of the Notch receptor is pulled away from the cell surface. Two proteolytic cleavages then occur; the first cleavage in a site exposed by the removal of the extracellular domain caused by ADAM metalloproteinases and the second cleavage at an intracellular site by γ-secretase. These cleavages result in the translocation of the NICD to the nucleus. Here it binds to a CSL protein complex to form a transcriptional co-activator. Transcription of downstream target genes such as those of the Hes and Hey families occurs. **ADAM**, a disintegrin and metalloproteinase

and to inhibit signalling (Hicks 2002). When the receptor binds one of its ligands, the NECD is pulled away from the intracellular domain (NICD) and is endocystosed by the ligand-bearing cell (Nichols et al. 2007). The removal of the NECD exposes a site within the transmembrane domain at which a proteolytic cleavage can occur by members of the ADAM (a disintegrin and metalloproteinases) protease family. A second cleavage by gamma-secretase, (Liu et al. 2004), which takes place at an intracellular position, results in the release of the NICD and its subsequent translocation to the nucleus. Here NICD is known to interact with members of the CSL (CBF1, suppressor of hairless, LAG2) superfamily (de la Pompa et al. 1997). The NICD/CSL interaction disrupts a CSL co-repressor complex and replaces it with transcription-activating proteins. The activation of CSL results in the transcription of a number of target genes in the Hes and Hey superfamilies that play major roles in cell fate determination and regulation of differentiation (Ross et al. 2004).

Recently, the Notch signalling pathway has been shown to be active in various stages of breast development; it has been shown to regulate both asymmetric division and SC self-renewal during early development and also to bias progenitor cell fate towards myoepithelial cell types (Woodward et al. 1998). The pathway is known to be crucial in normal breast development, and the Notch receptors, ligands and downstream target genes can be seen in the luminal cells of the normal human breast (Stylianou et al. 2006). Deregulation of expression and activation of the pathway have been observed in cancer and contribute to the transformed phenotype (Reedijk et al. 2005; Stylianou et al. 2006). Positive immunohistochemical staining for NICD expression was associated with a shorter median time to recurrence at 60 months after surgery for ductal carcinoma in situ (DCIS; Farnie et al. 2007).

The expression of Msi1 in the SP cells shown by Clarke et al. (2005) supports the vital role of the Notch self-renewal pathway in breast development. Msi1 is known to inhibit the production of Numb, a protein that blocks signalling through the Delta/Notch pathway and, as a result, allows active signalling. The production of transit amplifying and more differentiated cells from SCs is likely to occur by asymmetric division, which is known to be regulated by a Delta/Notch signalling pathway (Imai et al. 2001). Signalling at the stem/transit boundary between stem and non-stem cells would mean that, when ligand binding occurs, the NICD translocates to the nucleus in those cells that are positive for Msi1 but not in the adjacent cell. This process may account for the asymmetric division and the maintenance, via self-renewal, of the SC compartment.

In vitro culture has been used to study the effects of disrupted Notch signalling on human breast cells and to identify the role of Notch signalling in cell fate determination. One method, termed mammosphere culture, is used to show self-renewal and anoikis resistance of breast SCs. This culture system is based on neurosphere culture of neural SCs, which was established in the 1992 (Reynolds et al. 1992). Over-activation of Notch signalling by the addition of exogenous ligands has been shown to result in a ten-fold increase in sphere formation and increased myoepithelial cell formation and, therefore, a loss of the cellular hierarchy (Dontu et al. 2004). Inhibition of signalling, using a Notch 4 antibody or γ-secretase inhibitor, results in a decrease or complete lack of sphere formation in mammary cells taken from normal breast (Dontu et al. 2004) and pre-invasive DCIS (Farnie et al. 2007).

Another approach carried out to eliminate the effect Notch signalling was the production of a knockout mouse model of the downstream transcription factor RBP-J (Buono et al. 2006). This study shows a number of examples of loss of cell lineage organisation; proliferation is seen in basal cells rather the luminal cells, the luminal cells are seen to take on more basal characteristics and the level of steroid receptor-positive cells increases. Taken together, these results suggest that Notch plays an important role in both self-renewal and lineage differentiation.

Summary

The development of a complex structure such as the breast requires tightly controlled regulation in order for organogenesis to occur normally. Steroids and other local developmental signalling pathways such as Notch are clearly important factors in mammary gland development, regulating both the SCs that form the epithelium and their differentiation into functional, milk-producing and muscle-like cell types. Deregulation of their function is likely to result in arrested development and cancer.

References

Allan GJ, Beattie J, Flint DJ (2004) The role of IGFBP-5 in mammary gland development and involution. Domest Anim Endocrinol, 2004. 27:257–266

Alvi AJ, Clayton H, Joshi C, Enver T, Ashworth A, Vivanco MM, Dale TC, Smalley MJ (2003) Functional and molecular characterisation of mammary side population cells. Breast Cancer Res 5:R1–8

Asselin-Labat ML, Shackleton M, Stingl J, Vaillant F, Forrest NC, Eaves CJ, Visvader JE, Lindeman GJ (2006) Steroid hormone receptor status of mouse mammary stem cells. J Natl Cancer Inst 98:1011–1014

Bonnet D (2003) Hematopoietic stem cells. Birth Defects Res C Embryo Today 69:219–229

Brisken C, Park S, Vass T, Lydon JP, O'Malley BW, Weinberg RA (1998) A paracrine role for the epithelial progesterone receptor in mammary gland development. Proc Natl Acad Sci USA 95:5076–5081

Buono KD, Robinson GW, Martin C, Shi S, Stanley P, Tanigaki K, Honjo T, Hennighausen L (2006) The canonical Notch/RBP-J signaling pathway controls the balance of cell lineages in mammary epithelium during pregnancy. Dev Biol 293:565–580

Callahan R, Egan SE (2004) Notch signaling in mammary development and oncogenesis. J Mammary Gland Biol Neoplasia 9:145–163

Chepko G, Smith GH (1997) Three division-competent, structurally-distinct cell populations contribute to murine mammary epithelial renewal. Tissue Cell 29:239–253

Clarke RB, Howell A, Potten CS, Anderson E (1997) Dissociation between steroid receptor expression and cell proliferation in the human breast. Cancer Res 57:4987–4991

Clarke RB, Spence K, Anderson E, Howell A, Okano H, Potten CS (2005) A putative human breast stem cell population is enriched for steroid receptor-positive cells. Dev Biol 277:443–456

Clayton H, Titley I, Vivanco M (2004) Growth and differentiation of progenitor/stem cells derived from the human mammary gland. Exp Cell Res 297:444–460

Coleman S, Silberstein GB, Daniel CW (1988) Ductal morphogenesis in the mouse mammary gland: evidence supporting a role for epidermal growth factor. Dev Biol 127:304–315

Daniel CW, Young LJ (1971) Influence of cell division on an aging process. Life span of mouse mammary epithelium during serial propogation in vivo. Exp Cell Res 65:27–32

Daniel CW, Smith GH (1999) The mammary gland: a model for development. J Mammary Gland Biol Neoplasia 4:3–8

Daniel CW, Silberstein GB, Strickland P (1987) Direct action of 17 beta-estradiol on mouse mammary ducts analyzed by sustained release implants and steroid autoradiography. Cancer Res 47:6052–6057

de la Pompa JL, Wakeham A, Correia KM, Samper E, Brown S, Aguilera RJ, Nakano T, Honjo T, Mak TW, Rossant J, Conlon RA (1997) Conservation of the Notch signalling pathway in mammalian neurogenesis. Development 124:1139–1148

DeOme KB, Faulkin LJ Jr, Bern HA, Blair PB (1959) Development of mammary tumors from hyperplastic alveolar nodules transplanted into gland-free mammary fat pads of female C3H mice. Cancer Res19:515–520

Dontu G, Abdallah WM, Foley JM, Jackson KW, Clarke MF, Kawamura MJ, Wicha MS (2003) In vitro propagation and transcriptional profiling of human mammary stem/progenitor cells. Genes Dev 17:1253–1270

Dontu G, Jackson KW, McNicholas E, Kawamura MJ, Abdallah WM, Wicha MS (2004) Role of Notch signaling in cell-fate determination of human mammary stem/progenitor cells. Breast Cancer Res 6:R605–615

Ewan KB, Oketch-Rabah HA, Ravani SA, Shyamala G, Moses HL, Barcellos-Hoff MH (2005) Proliferation of estrogen receptor-alpha-positive mammary epithelial cells is restrained by transforming growth factor-beta1 in adult mice. Am J Pathol 167:409–417

Farnie G, Clarke RB, Spence K, Brennan KR, Anderson N, Bundred NJ (2007) Novel cell culture technique for primary ductal carcinoma in situ: Role of Notch and EGF receptor signaling pathways. J Natl Cancer Inst. 99:616–627

Furth PA, Bar-Peled U, Li M (1997) Apoptosis and mammary gland involution: reviewing the process. Apoptosis 2:19–24

Goodell MA, Rosenzweig M, Kim H, Marks DF, Demaria M, Paradis G, Grupp SA, Sieff CA, Mulligan RC, Johnson RP (1997) Dye efflux studies suggest that haematopoietic stem cells expressing low or undetectable levels of CD34 antigen exist in multiple species. Nature Med 3:1337–1345

Hicks C, Ladi E, Lindsell C, Hsieh JJ, Hayward SD, Collazo A, Weinmaster G (2002) A secreted Delta1-Fc fusion protein functions both as an activator and inhibitor of Notch1 signalling. J Neuro Res 68:655–667

Hovey RC, Trott JF, Vonderhaar BK (2002) Establishing a framework for the functional mammary gland: from endocrinology to morphology. J Mammary Gland Biol Neoplasia 7:17–38

Imai T, Tokunaga A, Yoshida T, Hashimoto M, Mikoshiba K, Weinmaster G, Nakafuku M, Okano H (2001) The neural RNA-binding protein Musashi1 translationally regulates mammalian numb gene expression by interacting with its mRNA. Mol Cell Biol 21:3888–3900

Jordan CT, Lemischka IR (1990) Clonal and systemic analysis of long-term hematopoiesis in the mouse. Genes Dev 4:220–232

Keeling JW, Ozer E, King G, Walker F (2000) Oestrogen receptor alpha in female fetal, infant, and child mammary tissue. J Pathol 191:449–451

Kopan R (2002) Notch: a membrane-bound transcription factor. J Cell Sci 115:1095–1097

Kordon EC, Smith GH (1998) An entire functional mammary gland may comprise the progeny from a single cell. Development 125:1921–1930

Lai EC (2004) Notch signaling: control of cell communication and cell fate. Development 131(5):965-973

Liu BY, McDermott SP, Khwaja SS, Alexander CM (2004) The transforming activity of Wnt effectors correlates with their ability to induce the accumulation of mammary progenitor cells. Proc Natl Acad Sci USA 101:4158–4163

Luetteke NC, Qiu TH, Fenton SE, Troyer KL, Riedel RF, Chang A, Lee DC (1999) Targeted inactivation of the EGF and amphiregulin genes reveals distinct roles for EGF receptor ligands in mouse mammary gland development. Development 126:2739–2750

Mallepell S, Krust A, Chambon P, Brisken C (2006) Paracrine signalling through the epithelial estorgen receptor alpha is required for proliferation and morphogenesis in the mammary gland. Proc Natl Acad Sci USA 103:2196–2201

Mayhall EA, Paffett-Lugassy N, Zon LI (2004) The clinical potential of stem cells. Curr Opin Cell Biol 16:713–720

Morgan T (1917) The theory of a gene. Am Nat 51:513–544

Morrison SJ (1997) Regulatory mechanisms in stem cell biology. Cell 88:287–298

Mumm JS, Schroeter EH, Saxena MT, Griesemer A, Tian X, Pan DJ, Ray WJ, Kopan R (2000) A ligand-induced extracellular cleavage regulates gamma-secretase-like proteolytic activation of Notch1. Mol Cell 5:197–206

Naccarato AG, Viacava P, Vignati S, Fanelli G, Bonadio AG, Montruccoli G, Bevilacqua G (2000) Bio-morphological events in the development of the human female mammary gland from fetal age to puberty. Virchows Arch 436:431–438

Nichols JT, Miyamoto A, Olsen SL, D'Souza B, Yao C, Weinmaster G (2007) DSL ligand endocytosis physically dissociates Notch1 heterodimers before activating proteolysis can occur. J Cell Biol 176:445–458

Peterson OW, Høyer PE, van Deurs B (1987) Frequency and distribution of estrogen receptor-positive cells in normal, nonlactating human breast tissue. Cancer Res 47:5748–5751

Politi K, Feirt N, Kitajewski J (2004) Notch in mammary gland development and breast cancer. Semin Cancer Biol 14:341–347

Potten CS, Loeffler M (1990) Stem cells: attributes, cycles, spirals, pitfalls and uncertainties. Lessons for and from the crypt. Development 110:1001–1020

Reedijk M, Odorcic S, Chang L, Zhang H, Miller N, McCready DR, Lockwood G, Egan SE (2005) High-level coexpression of JAG1 and NOTCH1 is observed in human breast cancer and is associated with poor overall survival. Cancer Res 65:8530–8537

Reynolds BA, Tetzlaff W, Weiss S (1992) A multipotent EGF-responsive striatal embryonic progenitor cell produces neurons and astrocytes. J Neurosci 12:4565–4574

Ross DA, Rao PK, Kadesch T (2004) Dual roles for the Notch target gene Hes-1 in the differentiation of 3T3-L1 preadipocytes. Mol Cell Biol 24:3505–3513

Shackleton M, Vaillant F, Simpson KJ, Stingl J, Smyth GK, Asselin-Labat ML, Wu L, Lindeman GJ, Visvader JE (2006) Generation of a functional mammary gland from a single stem cell. Nature 439:84–88

Siminovitch L, McCulloch EA, Till JE (1963) The distribution of colony-forming cells among spleen colonies. J Cell Physio 62:327–336

Smith GH, Medina D (1988) A morphologically distinct candidate for an epithelial stem cell in mouse mammary gland. J Cell Sci 90:173–183

Smith CA, Monaghan P, Neville AM (1984) Basal clear cells of the normal breast. Virchows Arch A Pathol Anat Histopathol 402:319–329

Smith GH, Strickland P, Daniel CW, Daniel CW (2002) Putative epithelial stem cell loss corresponds with mammary growth senescence. Cell Tissue Res 310:313–320

Stingl J, Eirew P, Ricketson I, Shackleton M, Vaillant F, Choi D, Li HI, Eaves CJ (2006) Purification and unique properties of mammary epithelial stem cells. Nature 439:993–997

Stylianou S, Clarke RB, Brennan K (2006) Aberrant activation of notch signaling in human breast cancer. Cancer Res 66:1517–1525

Vogel PM, Georgiade NG, Fetter BF, Vogel FS, McCarty KS (1981) The correlation of histologic changes in the human breast with the menstrual cycle. Am J Pathol 104:23–34

Welm BE, Tepera SB, Venezia T, Graubert TA, Rosen JM, Goodell MA (2002) Sca-1 (pos) cells in the mouse mammary gland represent an enriched progenitor cell population. Dev Biol 245:42–56

Welm BE, Freeman KW, Chen M, Contreras A, Spencer DM, Rosen JM (2002) Inducible dimerization of FGFR1: development of a mouse model to analyze progressive transformation of the mammary gland. J Cell Biol 157:703–714

Wilson CL, Sims AH, Howell A, Miller CJ, Clarke RB (2006) Effects of oestrogen on gene expression in epithelium and stroma of normal human breast tissue. Endocr Relat Cancer 13:617–628

Woodward TL, Xie, JW, Haslam SZ (1998) The role of mammary stroma in modulating the proliferative response to ovarian hormones in the normal mammary gland. J Mammary Gland Biol Neoplasia 3:117–131

Zeps N, Bentel JM, Papadimitriou JM, D'Antuono MF, Dawkins HJ (1998) Estrogen receptor-negative epithelial cells in mouse mammary gland development and growth. Differentiation 62:221–226

Estrogens, Cell Proliferation and Breast Cancer

Robert L. Sutherland[1], *C. Marcelo Sergio*[1], *Catriona M. McNeil*[1], *Luke R. Anderson*[1], *Claire K. Inman*[1], *Alison J. Butt*[1], and *Elizabeth A. Musgrove*[1]

Summary

Estrogens are potent mitogens in some target tissues including the mammary gland, where they play a pivotal role in normal development and in the initiation and progression of mammary carcinoma. The demonstration that estrogen-induced mitogenesis is associated with an increased rate of progression through G_1 phase of the cell cycle has focused attention on the estrogen regulation of molecules that control G_1 to S phase progression. Steroid-responsive breast cancer cells pretreated with pure estrogen antagonists arrest in G_0 and respond to estrogen with synchronous progression into S phase. This process is preceded by increased expression of c-Myc and cyclin D1, with consequent activation of cyclin D1-Cdk4 and cyclin E-Cdk2 complexes. These processes are mimicked by the inducible expression of either c-Myc or cyclin D1. Conversely, inhibition of estrogen-induced c-Myc gene expression using antisense or siRNA blocks S phase entry. Both cyclin D1 and c-Myc are mammary epithelial oncogenes that are frequently overexpressed in human breast cancers, and their ectopic expression in ER+ breast cancer cell lines decreases sensitivity to estrogen antagonists, implicating these key cell cycle regulatory molecules in the acquisition of endocrine resistance. In an attempt to identify targets of estrogen and c-Myc action with likely roles in proliferation control and antiestrogen resistance, Affymetrix oligonucleotide arrays were employed to compare transcript profiles of antiestrogen-arrested cells stimulated to re-initiate cell cycle progression by either estrogen treatment or c-Myc induction. After 6 hours of treatment, $\sim 50\%$ of estrogen-regulated genes were also regulated by c-Myc. Genes involved in cell growth (ribosome biogenesis, protein synthesis), cell proliferation (cell cycle control, DNA synthesis) and apoptosis were significantly over-represented in the estrogen-regulated gene set. These data indicate that a significant component of estrogen-induced growth and mitogenesis is mediated as a consequence of estrogen induction of c-Myc.

Introduction

Sex steroid hormones play a major role in the growth and development of estrogen target tissues, including the mammary gland, where they interact with other

---

[1] Cancer Research Program, Garvan Institute of Medical Research, St Vincent's Hospital, Sydney, NSW 2010, Australia

Address to correspondence: Cancer Research Program, Garvan Institute of Medical Research, 384 Victoria St, Darlinghurst NSW 2010, Australia, *email*: r.sutherland@garvan.org.au

---

Melmed et al.
Hormonal Control of Cell Cycle
© Springer-Verlag Berlin Heidelberg 2008

hormones, growth factors and cytokines in the precise regulation of cell proliferation, cellular differentiation and cell death. The classical model of 17β-estradiol (E$_2$) action involves ligand-mediated activation of the nuclear estrogen receptors, ERα and ERβ, which interact either directly with estrogen response elements (ERE) in the promoters of target genes or indirectly through protein-protein interactions with other transcription factors, including Sp1, c-Jun/ATF-2 and AP-1, to recruit coactivators and other components of the basic transcriptional machinery essential for transcriptional regulation (McDonnell and Norris 2002). Increasing evidence indicates that E$_2$ also induces acute nongenomic effects via signaling pathways more commonly associated with growth factor activation of cell surface receptors (Edwards 2005; Wong et al. 2002), e.g., the Src family of non-receptor tyrosine kinases and MAP kinases, to activate other transcriptional regulators. These signaling cascades complement the directly ER-regulated pathways to contribute to the global transcriptional response to estrogen and consequent altered cellular function.

The cell cycle phase-specific effects of estrogens on cell proliferation (Sutherland et al. 1983a, 1988) have focused attention on the role of estrogens and their receptors in regulating processes controlling the entry into, progression through, and exit from the G$_1$ phase of the cell cycle. Central to these cell cycle control mechanisms are cyclin D1-Cdk4 and cyclin E-Cdk2, which phosphorylate substrates including the product of the retinoblastoma susceptibility gene, pRB, thereby allowing initiation of DNA synthesis (Weinberg 1995). These enzyme complexes are controlled by several mechanisms, including transcriptional activation of cyclin gene expression; regulatory phosphorylation/dephosphorylation of the cyclin dependent kinase (CDK) subunits by kinases/phosphatases, including the CDK-activating kinase (CAK) and the Cdc25 family of phosphatases, and interactions with members of two distinct families of CDK inhibitors of which p16^{INK4A} and p21$^{WAF1/CIP1}$ are prototypic (Hunter and Pines 1994; Sherr and Roberts 1999). Essentially all of these modes of regulation have been documented following estrogen treatment (Foster et al. 2001a; Doisneau-Sixou et al. 2003).

Another well-studied target of estrogen action is c-Myc, a nuclear phosphoprotein of the basic helix-loop-helix family of transcription factors (Oster et al. 2002; Eisenman 2002), the expression/function of which is rate-limiting for progression through G$_1$ phase of the cell cycle (Hanson et al. 1994). Part of this action is due to the well-documented effects of c-Myc on activation of the cyclin E-Cdk2 complex (Rudolph et al. 1996). Furthermore, inducible expression of c-Myc re-initiates cell cycle progression in antiestrogen-arrested breast cancer cells (Prall et al. 1998), potentially implicating c-Myc in estrogen-induced mitogenesis and antiestrogen resistance. Thus, the recent expansion of knowledge about the molecular mechanisms regulating rates of cell cycle progression in steroid-responsive breast cancer cells has provided a framework within which to develop deeper insight into the mechanistic basis of estrogen-induced mitogenesis and antiestrogen action in breast cancer. Here we briefly summarize past work from this laboratory employing a "hypothesis testing" approach and preview ongoing research using an unbiased, genome-wide approach to the identification of estrogen-regulated genes involved in the control of cell growth, cellular proliferation and apoptosis in human breast cancer cells.

Mechanisms of Growth Inhibition by Antiestrogens

The initial observation that therapeutically relevant antiestrogenic drugs, e.g., tamoxifen, exerted their growth inhibitory effects on breast cancer cells via inhibition of cell cycle progression in G_1 phase of the cell cycle arose from in vitro experiments wherein ER-positive, estrogen-responsive, MCF-7 cells were arrested with tamoxifen (Sutherland et al. 1983b; Taylor et al. 1983). Subsequent development of pure antiestrogens, i.e., those with no estrogen agonist activity (Wakeling and Bowler 1988) in contrast with the partial agonist/partial antagonist properties of tamoxifen, and the elucidation of the roles of cyclins, Cdks and their inhibitors in mammalian cell cycle control (Hunter and Pines 1994) facilitated a deeper understanding of the molecular basis of antiestrogen-mediated cell cycle arrest.

The pure antiestrogen, ICI 182780, arrests MCF-7 cells in a quiescent (G_0) state characterized by the formation of p130/E2F4 complexes and the accumulation of hyperphosphorylated E2F4 (Fig. 1a; Carroll et al. 2000). In this state, cells are relatively resistant to the mitogenic effects of growth factors, as evidenced by a failure of insulin, IGF-1 and EGF to reinitiate cell cycle progression in the absence of estrogen (Lai et al. 2001). Antiestrogen-mediated arrest is associated with decreased cyclin D1 gene expression, inactivation of cyclin D1-Cdk4 complexes, and decreased phosphorylation of

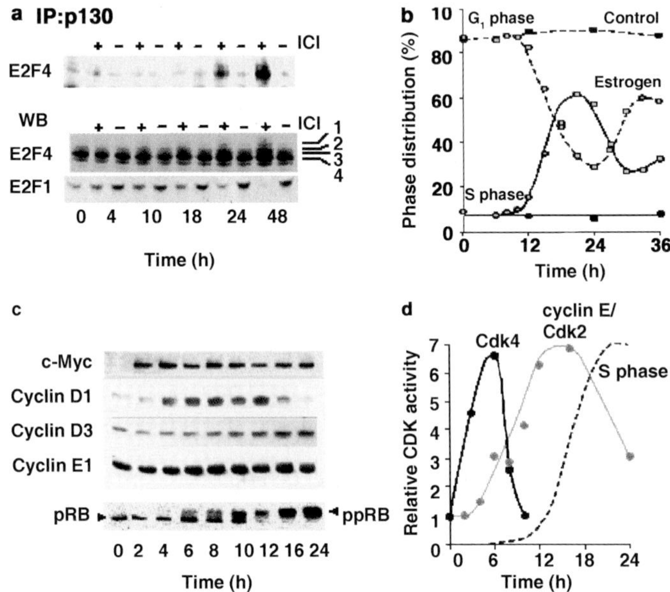

Fig. 1. Estrogen-induced cell cycle progression in MCF-7 cells pretreated with the pure antiestrogen ICI 182780. (**a**) Western blots of lysates from cells treated with ICI 182780 for 48 hrs demonstrating the formation of p130-E2F4 complexes and hyperphosphorylation of E2F4, which are markers of cellular quiescence (G_0). (**b**) Cell cycle phase distribution following estrogen treatment of antiestrogen-arrested cells. (**c**) Western blots of temporal changes in c-Myc, cyclins D1, D3 and E, and pRb phosphorylation following estrogen treatment. (**d**) Temporal changes in Cdk4 and cyclin E-Cdk2 activities and S phase entry following estrogen treatment

pRB (Musgrove et al. 1993; Watts et al. 1995). Inhibition of cyclin E-Cdk2 activity also occurs prior to a decrease in the S phase fraction and is dependent on p21$^{WAF1/CIP1}$, since treatment with antisense oligonucleotides to p21$^{WAF1/CIP1}$ attenuates this effect (Carroll et al. 2000). Recruitment of p21$^{WAF1/CIP1}$ to cyclin E-Cdk2 complexes is, in turn, dependent on decreased cyclin D1 expression and decreased sequestration of p21$^{WAF1/CIP1}$ into cyclin D1-Cdk4 complexes (Carroll et al. 2000). p27^{Kip1} is another essential mediator of cell cycle arrest by antiestrogens, as indicated by antisense-mediated down-regulation of p27^{Kip1} abrogating antiestrogen-induced cell cycle arrest in MCF-7 cells (Cariou et al. 2000). However, p21$^{WAF1/CIP1}$ antisense treatment resulted in a decrease in p27^{Kip1} protein levels, whereas changes in p21$^{WAF1/CIP1}$ protein levels were not observed following treatment with p27^{Kip1} antisense (Carroll et al. 2000). These data suggest that the synthesis and redistribution of p21$^{WAF1/CIP1}$ initiate the inhibition of the cyclin E-Cdk2 complex, with consequent accumulation of p27^{Kip1} further amplifying the inhibitory response.

Among the first candidate genes to be investigated as a potential target of estrogen/antiestrogen-regulated cell proliferation was the proto-oncogene, c-Myc. Rapid decreases in c-*myc* mRNA and c-Myc protein levels are observed in response to a spectrum of antiestrogens in both in vivo and in vitro models (Watts et al. 2002). Therefore, in addition to cyclin D1 and p21$^{WAF1/CIP1}$, c-Myc is a likely target molecule through which antiestrogens mediate cell cycle control and indeed may be the pivotal initiating event. Evidence in support of the latter concept arises from data from MCF-7 cells treated with antisense oligonucleotides to c-Myc. Such treatment leads to inhibition of cyclin D1 expression, subsequent redistribution of p21$^{WAF1/CIP1}$ from cyclin D1-Cdk4 to cyclin E-Cdk2 complexes, and a decline in cyclin E-Cdk2 enzymatic activity (Carroll et al. 2002), thereby recapitulating the initial downstream events that culminate in growth arrest after ICI 182780 treatment. Simultaneous repression of p21$^{WAF1/CIP1}$ attenuated the growth-inhibitory effects of reduced c-Myc expression, emphasizing the importance of this CDK inhibitor in c-Myc action in these cells (Carroll et al. 2002). The subsequent demonstration that p21$^{WAF1/CIP1}$ transcription is repressed by c-Myc (Gartel et al. 2001) provides a mechanistic link between antiestrogen-mediated down-regulation of c-Myc, increased p21$^{WAF1/CIP1}$ gene expression, increased p21$^{WAF1/CIP1}$ binding to cyclin E-Cdk2 complexes, decreased Cdk2 activity and cell cycle arrest.

Estrogen Control of Cell Cycle Progression

The reversible inhibition of antiestrogen-mediated cell cycle arrest in breast cancer cells by competitive replacement of antagonist, ICI 182780, with agonist, estradiol, at the ER ligand binding site has provided a robust model for elucidating the molecular events initiated early in the mitogenic response to estrogen (Prall et al. 1997). Following a lag of about 10 hours, cells arrested in G_0 have traversed G_1 phase and begin to enter S phase in a highly synchronous manner (Fig. 1b). Amongst the earliest detectable transcriptional events following addition of estradiol is the rapid induction of c-*myc* mRNA and c-Myc protein (Fig. 1c: Prall et al. 1997; Dubik et al. 1987). Strong evidence that c-Myc is likely to play a significant regulatory role in estrogen action is provided by the demonstration that c-Myc antisense oligonucleotides inhibit estrogen-stimulated breast cancer cell proliferation (Watson et al. 1991), a result recently confirmed with c-Myc siRNA in our "estrogen-rescue" model (Anderson et al., unpublished data).

This initial, acute activation of c-Myc is followed by pronounced increases in cyclin D1 mRNA and protein expression (Fig. 1c), cyclin D1-Cdk4 complex formation and augmented Cdk4 enzymatic activity (Prall et al. 1997; Altucci et al. 1996; Planas-Silva and Weinberg 1997). The increase in cyclin D1 expression occurs significantly later than the earliest changes in c-*myc* mRNA expression but coincides with increased phosphorylation of pRB and precedes S phase entry by some 9 hours (Fig. 1c, d; Prall et al. 1997). Although effects on protein stability are also possible, the effect of estrogen on cyclin D1 protein expression appears to be predominantly transcriptionally mediated, since increased expression of cyclin D1 mRNA precedes changes in cyclin D1 protein (Prall et al. 1997; Altucci et al. 1996). However, the cyclin D1 gene regulatory region does not contain a classical estrogen response element (ERE). Studies of the proximal promoter have identified a number of estrogen-responsive enhancer elements, including an atypical cAMP response element (CRE) and Sp1 sites close to the transcription start site, and more distal Sp1 sites (Altucci et al. 1996; Castro-Rivera et al. 2001; Cicatiello et al. 2004; Sabbah et al. 1999). Recent studies have identified a further two enhancer elements (Eeckhoute et al. 2006). One of these is located downstream of the cyclin D1 coding region and recruits the ER in a Fox A1-dependent manner. The chromatin-remodeling activity of FoxA1 then facilitates recruitment of other transcription factors, including Sp1 (Eeckhoute et al. 2006). Importantly, some factors interacting with this enhancer element are cell-specific, providing a potential explanation for the observation that ectopic expression of the ERα is not sufficient for estrogen regulation of cyclin D1 (Planas-Silva et al. 1999).

Compelling evidence that cyclin D1 plays an essential role in estrogen-induced cell cycle progression comes from studies in which cyclin D1 is inhibited functionally. Thus, when either antibodies against cyclin D1 or the Cdk4-specific inhibitor p16^{INK4A} are introduced into MCF-7 cells by microinjection, estrogen fails to stimulate G_1/S phase progression (Lukas et al. 1996), indicating that cyclin D1 is necessary for estrogen-induced cell cycle progression.

In addition to activation of cyclin D1-Cdk4 complexes, estrogen also activates cyclin E-Cdk2 complexes within 4 hours, substantially preceding entry into S phase (Fig. 1d; Prall et al. 1997; Foster and Wimalasena 1996; Planas-Silva and Weinberg 1997). This very early activation of cyclin E-Cdk2 following E_2 administration is in marked contrast to cell cycle progression stimulated by other growth factor mitogens, where cyclin E-Cdk2 activation is associated more with transcription activation of cyclin E1 closer to the G_1/S phase transition (Koff et al. 1992). This finding suggests that the early activation of cyclin E-Cdk2 has a particularly important role in estrogen-induced cell cycle progression. However, in the first hours following estrogen treatment, there is little change in protein levels of cyclin E1, Cdk2, or the CDK inhibitors, p21$^{WAF1/CIP1}$ and p27^{Kip1} in either total cell lysates or in cyclin E-Cdk2 complexes (Prall et al. 1997). The mechanistic basis for this effect was not initially apparent and was subsequently explored in detail by two laboratories (Prall et al. 1997, 2001; Planas-Silva and Weinberg 1997).

Separation of the cyclin E-Cdk2 complexes by gel filtration chromatography indicated that estrogen treatment was associated with the formation of high molecular weight complexes (Prall et al. 1997) and that induction of c-Myc or cyclin D1 likewise led to the formation of active high molecular weight cyclin E-Cdk2 complexes (Prall et al. 1998). These complexes constituted a minority of the cyclin E-Cdk2 protein but were of high specific activity, accounting for the majority of cyclin E-Cdk2 activity; they were also relatively deficient in p21$^{WAF1/CIP1}$ and p27^{Kip1}. Estrogen treatment relieved the inhibitory activity of p21$^{WAF1/CIP1}$ toward cyclin E-Cdk2 (Prall et al. 1997; Planas-Silva and Weinberg 1997), the result of a decrease in newly synthesized p21$^{WAF1/CIP1}$ (Prall et al. 2001) that in turn appears to result from transcriptional repression by estrogen-induced c-Myc, since p21$^{WAF1/CIP1}$ is a validated c-Myc-repressed target gene (Gartel et al. 2001). Subsequently, the pRB-related protein p130, which can compete with p21$^{WAF1/CIP1}$ for cyclin-CDK binding (Zhu et al. 1995; Shiyanov et al. 1996), is recruited to the cyclin E-Cdk2 complex following estrogen treatment and c-Myc or cyclin D1 induction, contributing to the increased size of the active complex (Prall et al. 1998). Whether the presence of p130 in the active cyclin E-Cdk2 complex is of functional significance or merely the result of its high abundance in ICI 182780-treated cells (Carroll et al. 2000) remains to be elucidated.

The more recent identification of a second mammalian cyclin E, cyclin E2 (Gudas et al. 1999), and the demonstration that it is markedly induced at both the mRNA and protein levels following estrogen stimulation (Musgrove et al., unpublished data) calls for a re-evaluation of the mechanisms of activation of Cdk2 by estrogen described above. An alternate mechanism could be via estrogen-mediated transcriptional activation of cyclin E2 and the consequent formation of active cyclin E2-Cdk2 complexes. This mechanism is currently under investigation.

Finally, a role for the Cdc25A phosphatase in the further activation of cyclin E-Cdk2 complexes is suggested from data demonstrating that antisense Cdc25A oligonucleotides inhibited estrogen-induced Cdk2 activation and DNA synthesis, whereas inactive cyclin E-Cdk2 complexes from p16^{INK4A}-expressing, estrogen-treated cells were activated in vitro by treatment with recombinant Cdc25A and in vivo in cells overexpressing Cdc25A (Foster et al. 2001b). These studies establish Cdc25A as another growth-promoting target of estrogen action and further indicate that estrogens independently regulate multiple components of the cell cycle machinery to facilitate progression from G_0 to S phase (Foster et al. 2001a; Doisneau-Sixou et al. 2003).

Cell Cycle Genes as Mediators of Antiestrogen Resistance

Several of the cell cycle targets of estrogen action in breast cancer cells are aberrantly expressed in human breast cancer. These were initially identified as gene amplification at the 8q24 (*MYC*) and 11q13 (*CCND1*) loci, but expression studies at the mRNA and protein level demonstrated that overexpression of c-Myc, cyclin D1 and cyclin E1 was more frequent than could be accounted for by increased gene copy number and amongst the most common aberrations in primary breast cancer (Table 1; Buckley et al. 1993; Wang and Shao 2006). These observations, in association with the demonstration that inducible expression of c-Myc and cyclin D1 could reverse the growth inhibitory effects of antiestrogens (Prall et al. 1998; Wilcken et al. 1997), led us to test the effects of overexpression of these genes on antiestrogen sensitivity in vitro.

The ability of the pure antiestrogen, ICI 182780, to decrease proliferation of MCF-7 cells expressing different levels of c-Myc was assessed in two experimental systems,

Table 1. Aberrations of cell cycle regulators in breast cancer

	Frequency range (%)	Mean (%)
MYC amplification	4–52	19
c-Myc overexpression	11–70	38
11q13 amplification	9–17	13
Cyclin D1 overexpression	28–81	45
Cyclin E1 overexpression	28–35	32
Decreased p27	50–63	57

using as an endpoint the percentage of the cell population in S phase 24–48 hours after antiestrogen treatment. Figure 2a demonstrates the attenuation of the antiproliferative effect of ICI 182780 in two clones of MCF-7 cells in which c-Myc was constitutively overexpressed. These data demonstrated that a $\sim$2-fold increase in c-Myc resulted in a $\sim$50% decrease in sensitivity to the antiproliferative effect of ICI 182780. To establish if higher levels of c-Myc overexpression could further attenuate these antiproliferative effects, c-Myc overexpression was induced in two clonal cell lines expressing c-Myc under the influence of a zinc-inducible promoter. c-Myc expression was rapidly induced following zinc treatment and, after 3 hours, the cells were treated with 10 nM ICI 182780. After 24 hours of treatment, cells with no c-Myc induction (i.e., 0 µM zinc added) demonstrated the expected decline in S phase to about 20% of that observed in untreated exponentially growing cells (Fig. 2b). However, incremental increases in c-Myc expression, as the concentration of zinc was increased, resulted in a concurrent incremental reduction in the ability of antiestrogen treatment to reduce the S fraction (Fig. 2b). At the highest concentration of zinc (60 µM), the antiestrogen-induced decline in S phase was significantly attenuated to $\sim$80% of that seen in untreated cells (Fig. 2b). Together these data clearly demonstrate that c-Myc overexpression can dampen the growth inhibitory response to antiestrogen in a concentration-dependent manner and potentially, at even higher levels of expression, render cells completely insensitive to antiestrogens. These and other published data (Venditti et al. 2002; Mukherjee and Conrad 2005) further support a potential role for c-Myc overexpression in the development of endocrine resistance in breast cancer.

A similar approach demonstrated that constitutive and inducible expression of cyclin D1 can rescue breast cancer cells from antiestrogen-induced growth arrest (Prall et al. 1998; Wilcken et al. 1997; Hodges et al. 2003). Constitutive overexpression of cyclin D1 in T-47D breast cancer cells decreased sensitivity to ICI 182780 in the short-term, up to 48 hours, but this decrease was less apparent by 72 hours post treatment (Fig. 2c), suggesting that cyclin D1 may abrogate the early cell cycle effects of antiestrogen inhibition (Hui et al. 2002). However, in spite of the apparent transient nature of this response, clonogenic survival assays demonstrated a $\sim$2-fold decreased sensitivity to antiestrogens in cyclin D1-overexpressing cells (Fig. 2d). Sustained expression of cyclin D1 is also seen in breast cancer cells during acquisition of tamoxifen resistance (Kilker et al. 2004). In these cells, ER expression and function remained intact, and the pure antiestrogen, ICI 164384, retained its anti-proliferative effects via suppression of cyclin D1. This finding is consistent with the clinically observed benefit seen in patients with tamoxifen-resistant breast cancers who respond to second line therapy with ER down-regulating, pure antiestrogens (Howell et al. 2005). Interestingly, overexpression

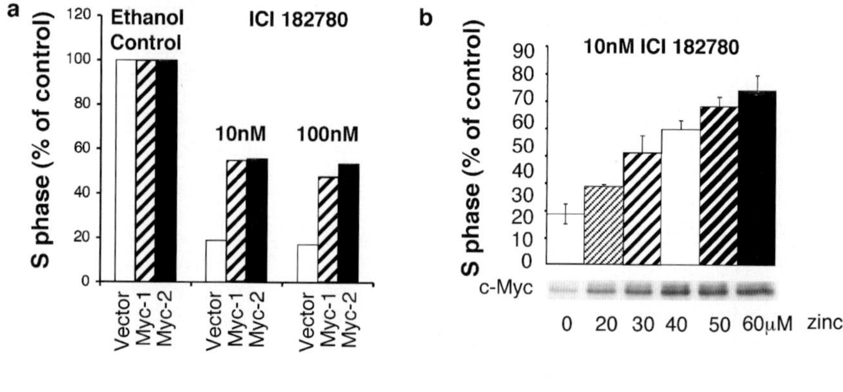

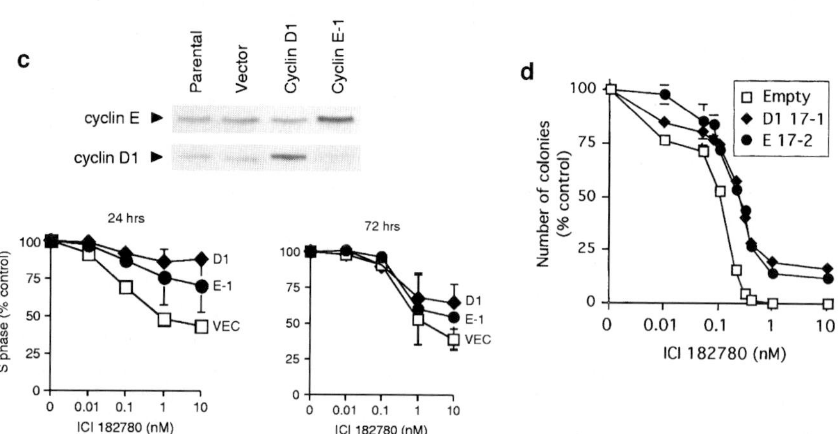

Fig. 2. Overexpression of either c-Myc, cyclin D1 or cyclin E in breast cancer cells attenuates the growth inhibitory effects of antiestrogen. (**a**) Two independent clones of MCF-7 cells constitutively overexpressing c-Myc $\sim$2-fold were treated with ICI 182780. S phase was measured 48 hrs later and expressed as a percentage of S phase in ethanol-treated control cells. (**b**) MCF-7 cells expressing c-Myc under the control of the metallothionen promoter were pretreated with increasing concentrations of zinc to induce c-Myc (*lower panel*), then treated with 10 nM ICI 182780 for 24 hrs. S phase was determined by flow cytometry and expressed as a percentage of S phase in vehicle-treated exponentially growing cells. (**c**) T-47D cells constitutively overexpressing cyclin D1 or cyclin E (*upper panel*) were treated with increasing concentrations of ICI 182780 and S phase was measured at 24 and 72 hrs. (**d**) Clonogenic survival assays of cyclin D1 and cyclin E overexpressing T-47D cells showing an $\sim$2- to 2.5-fold decrease in sensitivity to ICI 182780 compared to empty vector control cells. Reproduced from Hui et al. 2002

of cyclin D1 confers complete resistance to the growth inhibitory effects of progestins (Musgrove et al. 2001).

There also exist in vitro data supporting a role for cyclin E1 in the development of antiestrogen resistance. Studies in MCF-7 cells showed that a 3-fold overexpression of cyclin E1 abrogated tamoxifen-mediated growth arrest and also demonstrated that cyclin E1 overexpression confers partial resistance to the acute inhibitory effects of ICI 182780, although to a lesser extent than that observed with cyclin D1 (Fig. 2c,d: Hui et al. 2002; Dhillon and Mudryj 2002). Nonetheless, in clonogenic survival assays, overexpression of cyclin E1, like cyclin D1, conferred significant resistance to the growth inhibitory effects of ICI 182780 (Fig. 2d). The production of low-molecular weight isoforms of cyclin E, which appear unique to tumor cells, also confers resistance to antiestrogens in MCF-7 cells (Akli et al. 2004). Together these data provide compelling evidence that c-Myc, cyclin D1 and cyclin E1 are commonly overexpressed in primary breast cancers and, when overexpressed, can induce partial resistance to the growth inhibitory effects of several endocrine therapies in vitro. However, further definition of their potential roles in the acquisition of endocrine resistance in the clinical setting must await data from large, randomized treatment trials with concurrent assessment of tumor c-Myc, cyclin D1 and cyclin E status.

A Genome-wide Approach to the Identification of Estrogen Target Genes Involved in Cell Growth, Proliferation and Apoptosis

In an attempt to further identify estrogen-regulated genes that are transcriptionally regulated as a consequence of estrogen-induction of c-Myc, and that might potentially contribute to endocrine resistance mediated by c-Myc, a series of clonal MCF-7 cell lines was developed that expressed wild-type c-Myc or c-Zip (a deletion mutant lacking the N-terminal transactivation domain) under the control of the zinc-inducible metallothionein promoter (Fig. 3). Representative clones with estrogen and antiestrogen responses matched to those of the parental MCF-7 cells were chosen for further experiments. Zinc treatment resulted in increased c-Myc or c-Zip expression within 3 hours (Fig. 3a), similar to the timing of estrogen induction of c-Myc and consistent with our previous data using this zinc-inducible construct (Prall et al. 1998). The majority of the cell population synchronously re-initiated cell cycle progression following E_2 treatment. Induction of c-Myc also led to re-initiation of cell cycle progression to a degree comparable with the effects of estrogen. However, although cells transfected with c-Zip could respond to estrogen treatment, c-Zip induction did not lead to cell cycle progression (Fig. 3b).

RNA for transcript profiling was collected 6 hours after estrogen or zinc treatment, within the time-frame during which critical estrogen-dependent events necessary for S phase entry occur (Musgrove et al. 1989). Transcript profiling was undertaken in triplicate following estrogen treatment (compared with vehicle treatment) and c-Myc or c-Zip induction (compared with zinc-treated empty vector cells), using Affymetrix HG-U133 plus 2.0 arrays. Analysis of the microarray data used Bayesian linear modelling methods in the *limma* package and the Benjamini and Yuketieli adjustment was applied for multiple-hypothesis comparisons (Smyth et al. 2005; Benjamini et al. 2001). Probe sets that were significantly up- or down-regulated following estrogen treatment compared with vehicle-treated cells were identified (adjusted p < 0.01, Fig. 3c). These estrogen-regulated probe sets were further divided into those that were significantly regulated following c-Myc but not c-Zip induction, designated "E2 and Myc," and the remainder, which we designated "E2 not Myc". Approximately two-thirds of the

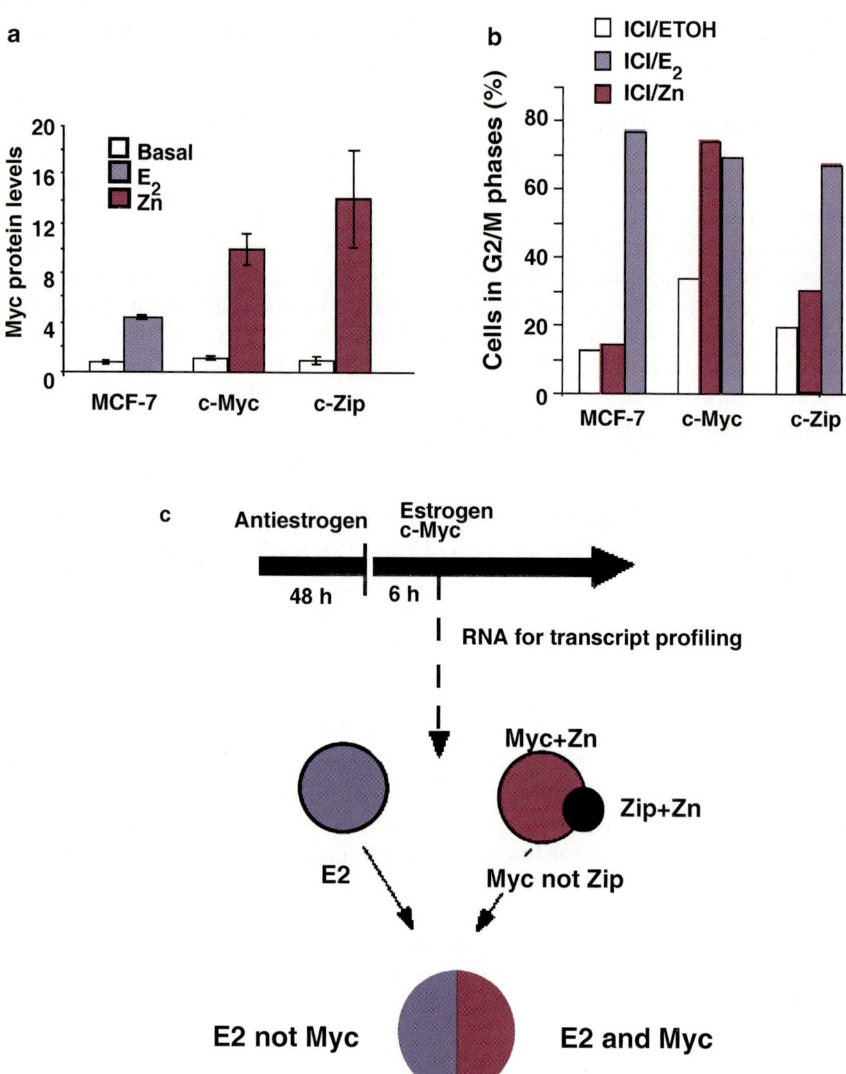

Fig. 3. Experimental model for identifying estrogen-regulated c-Myc target genes in breast cancer cells. MCF-7 cells inducibly expressing c-Myc (wild type or c-Zip), parental cells, and cells bearing the empty vector were pretreated with 10 nM ICI 182780 for 48 hrs. Parental cells were then treated with estrogen (100 nM E_2) or vehicle (ethanol), and cells transfected with c-Myc, c-Zip or the empty vector were treated with zinc (65 M). (**a**) Quantitation of Western blots of c-Myc levels 6 hrs after estrogen treatment or c-Myc induction in representative MCF-7 clones. (**b**) Cells were additionally treated with nocodazole to block estrogen- or c-Myc-stimulated cells in G2/M. Cell cycle phase distribution was determined after 36 hrs of estrogen or c-Myc/c-Zip induction using flow cytometry. (**c**) RNA was prepared from cells harvested 6 hrs following treatment for transcript profiling. The estrogen-regulated probe sets were compared with those regulated by c-Myc but not c-Zip to identify probe sets regulated by both estrogen and c-Myc (but not c-Zip), designated (E2 and Myc) or by estrogen but not c-Myc or c-Zip (E2 not Myc)

estrogen-regulated genes were upregulated and, in total, half of the probe sets significantly regulated by estrogen in this model system were also significantly regulated by c-Myc.

Publicly available databases of estrogen-responsive genes (ERGDB; Tang et al. 2004) and c-Myc targets (Zeller et al. 2003) were searched to determine what proportion of the probe sets in each category had been previously identified as either estrogen- or c-Myc-regulated. This analysis revealed that overall 29% of probe sets significantly upregulated by estrogen were present in the ERGDB, and 27% of those that were also significantly upregulated by c-Myc induction were present in the Myc target gene database. However, relatively few (12–14%) of the probe sets down-regulated in this experimental model were previously described estrogen or c-Myc targets.

The probe sets from the "E2 not Myc" category that increased in expression had the highest proportion of previously-documented estrogen targets, i.e. 39%. Within the estrogen-upregulated probe sets, 118/188 (63%) of the known estrogen targets were regulated by estrogen but not c-Myc, consistent with our initial premise that genes regulated by both estrogen and c-Myc have been under-represented in previous studies. Although the latter category contains relatively few previously identified estrogen targets, it likely contains a significant number of bona fide targets, since a high proportion of the probe sets in the "E2 and Myc" category that increased in expression was present in the Myc target gene database (39%).

Finally, in an initial attempt to identify biological processes that were significantly over-represented in the probe sets regulated by estrogen, we employed the data-mining tool Onto-Express (http://vortex.cs.wayne.edu/index.htm). These data are summarized in Fig. 4 and illustrate that the major functional categories of the estrogen-regulated genes include cell growth (rRNA processing, ribosome biogenesis, tRNA processing and protein synthesis), cell proliferation (DNA replication and regulation of CDK activity, cell cycle and cell proliferation) and apoptosis. Of interest and potential functional significance is the disparity in the proportion of "E2 not Myc" and "E2 and Myc" regulated genes in the different functional groups. For example, it appears that those genes involved in the estrogen regulation of cell growth are predominantly c-Myc target genes whereas genes involved in the control of the cell cycle and cell proliferation contain a mixture of "E2 not Myc" and "E2 and Myc" targets.

Conclusions

Research conducted over the past decade has facilitated a deeper understanding of the molecular basis of estrogen-induced mitogenesis, a major contributor to the development and progression of breast cancer. Concurrent studies identifying molecular aberrations associated with the development of the clinical disease and their functional characterization in animal models revealed several estrogen-regulated genes that are both overexpressed in breast cancer and oncogenic when expressed in mammary epithelial cells. The best characterized of these are c-Myc and cyclins D1 and E1. Furthermore, overexpression of these genes in vitro leads to decreased sensitivity to clinically relevant antiestrogenic therapies, i.e., tamoxifen and ICI 182780, implying

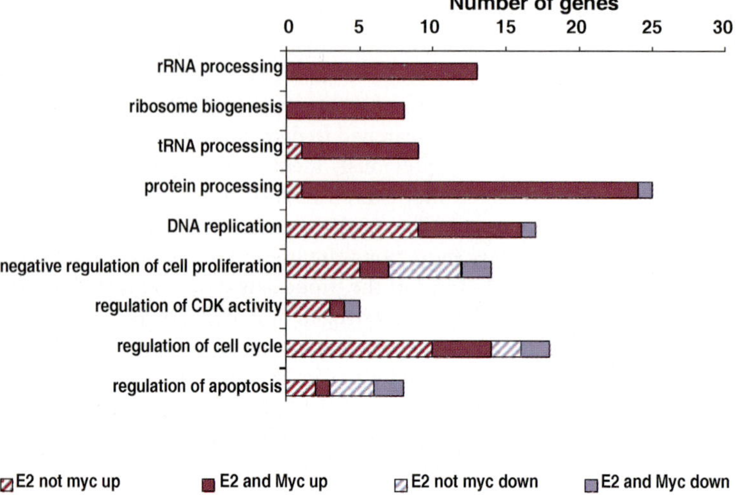

Fig. 4. Gene ontology of estrogen-regulated genes identified in the experiments outlined in Fig. 3. Onto-Express (http://vortex.cs.wayne.edu/index.htm) was used to identify biological processes that were significantly over-represented in the estrogen-regulated probe set on the Affymetrix chips. Genes regulated by "E2 and Myc" are represented by solid bars and "E2 not Myc" by cross-hatched bars and varied in their proportions between different functional categories

that they may play a role in the acquisition of endocrine resistance, a major limitation to the effective treatment of hormone-responsive breast cancer.

A major ongoing challenge is identifying other estrogen target genes involved in the etiology of breast cancer and its responsiveness to therapeutic intervention. We have described one experimental approach that has further elucidated the importance of c-Myc in the global estrogen response. These new data may identify previously unreported estrogen target genes with potential roles as markers of endocrine resistant disease or new molecular targets for therapeutic intervention.

Acknowledgements. Work in this laboratory is supported by grants from the following organizations: National Health and Medical Research Council of Australia (NH&MRC), Cancer Institute NSW, R.T. Hall Trust, Australian Cancer Research Foundation (ACRF) and Association for International Cancer Research (AICR). Catriona McNeil and Luke Anderson are Cancer Institute NSW Scholars; Catriona McNeil is also the recipient of an NH&MRC Postgraduate Scholarship. Elizabeth Musgrove and Alison Butt are Cancer Institute NSW Research Fellows. Rob Sutherland is a Senior Principal Research Fellow of the NH&MRC. We thank Chehani Alles and Margaret Gardiner-Garden for assistance with data analysis and Joanne Scorer for preparation of the manuscript.

References

Akli S, Zheng PJ, Multani AS, Wingate HF, Pathak S, Zhang N, Tucker SL, Chang S, Keyomarsi K (2004) Tumor-specific low molecular weight forms of cyclin E induce genomic instability and resistance to p21, p27, and antiestrogens in breast cancer. Cancer Res 64:3198–3208

Altucci L, Addeo R, Cicatiello L, Dauvois S, Parker MG, Truss M, Beato M, Sica V, Bresciani F, Weisz A (1996) 17beta-Estradiol induces cyclin D1 gene transcription, p36D1-p34cdk4 complex activation and p105Rb phosphorylation during mitogenic stimulation of G(1)-arrested human breast cancer cells. Oncogene 12:2315–2324

Benjamini Y, Drai D, Elmer G, Kafkafi N, Golani I (2001) Controlling the false discovery rate in behavior genetics research. Behav Brain Res 125:279–284

Buckley MF, Sweeney KJ, Hamilton JA, Sini RL, Manning DL, Nicholson RI, de Fazio A, Watts CK, Musgrove EA, Sutherland RL (1993) Expression and amplification of cyclin genes in human breast cancer. Oncogene 8:2127–2133

Cariou S, Donovan JC, Flanagan WM, Milic A, Bhattacharya N, Slingerland JM (2000) Down-regulation of p21WAF1/CIP1 or p27Kip1 abrogates antiestrogen-mediated cell cycle arrest in human breast cancer cells. Proc Natl Acad Sci USA 97:9042–9046

Carroll JS, Prall OW, Musgrove EA, Sutherland RL (2000) A pure estrogen antagonist inhibits cyclin E-Cdk2 activity in MCF-7 breast cancer cells and induces accumulation of p130-E2F4 complexes characteristic of quiescence. J Biol Chem 275:38221–38229

Carroll JS, Swarbrick A, Musgrove EA, Sutherland RL (2002) Mechanisms of growth arrest by c-*myc* antisense oligonucleotides in MCF-7 breast cancer cells: implications for the antiproliferative effects of antiestrogens. Cancer Res 62:3126–3131

Castro-Rivera E, Samudio I, Safe S (2001) Estrogen regulation of cyclin D1 gene expression in ZR-75 breast cancer cells involves multiple enhancer elements. J Biol Chem 276:30853–30861

Cicatiello L, Addeo R, Sasso A, Altucci L, Petrizzi VB, Borgo R, Cancemi M, Caporali S, Caristi S, Scafoglio C, Teti D, Bresciani F, Perillo B, Weisz A (2004) Estrogens and progesterone promote persistent CCND1 gene activation during G1 by inducing transcriptional derepression via c-Jun/c-Fos/estrogen receptor (progesterone receptor) complex assembly to a distal regulatory element and recruitment of cyclin D1 to its own gene promoter. Mol Cell Biol 24:7260–7274

Deming SL, Nass SJ, Dickson RB, Trock BJ (2000) C-myc amplification in breast cancer: a meta-analysis of its occurrence and prognostic relevance. Br J Cancer 83:1688–1695

Dhillon NK, Mudryj M (2002) Ectopic expression of cyclin E in estrogen responsive cells abrogates antiestrogen mediated growth arrest. Oncogene 21:4626–4634

Doisneau-Sixou SF, Sergio CM, Carroll JS, Hui R, Musgrove EA, Sutherland RL (2003) Estrogen and antiestrogen regulation of cell cycle progression in breast cancer cells. Endocr Relat Cancer 10:179–186

Dubik D, Dembinski TC, Shiu RP (1987) Stimulation of c-myc oncogene expression associated with estrogen-induced proliferation of human breast cancer cells. Cancer Res 47:6517–6521

Edwards DP (2005) Regulation of signal transduction pathways by estrogen and progesterone. Annu Rev Physiol 67:335–376

Eeckhoute J, Carroll JS, Geistlinger TR, Torres-Arzayus MI, Brown M (2006) A cell-type-specific transcriptional network required for estrogen regulation of cyclin D1 and cell cycle progression in breast cancer. Genes Dev 20:2513–2526

Eisenman R (2002) Deconstructing Myc. Genes Dev 15:2023–2030

Foster JS, Wimalasena J (1996) Estrogen regulates activity of cyclin-dependent kinases and retinoblastoma protein phosphorylation in breast cancer cells. Mol Endocrinol 10:488–498

Foster JS, Henley DC, Ahamed S, Wimalasena J (2001) Estrogens and cell-cycle regulation in breast cancer. Trends Endocrinol Metab 12:320–327

Foster JS, Henley DC, Bukovsky A, Seth P, Wimalasena J (2001) Multifaceted regulation of cell cycle progression by estrogen: regulation of Cdk inhibitors and Cdc25A independent of cyclin D1-Cdk4 function. Mol Cell Biol 21:794–810

136   R.L. Sutherland et al.

Gartel AL, Ye X, Goufman E, Shianov P, Hay N, Najmabadi F, Tyner AL (2001) Myc represses the p21(WAF1/CIP1) promoter and interacts with Sp1/Sp3. Proc Natl Acad Sci USA 98:4510–4515

Gudas JM, Payton M, Thukral S, Chen E, Bass M, Robinson MO, Coats S (1999) Cyclin E2, a novel G_1 cyclin that binds Cdk2 and is aberrantly expressed in human cancers. Mol Cell Biol 19:612–622

Hanson KD, Shichiri M, Follansbee MR, Sedivy JM (1994) Effects of c-myc expression on cell cycle progression. Mol Cell Biol 14:5748–5755

Hodges LC, Cook JD, Lobenhofer EK, Li L, Bennett L, Bushel PR, Aldaz CM, Afshari CA, Walker CL (2003) Tamoxifen functions as a molecular agonist inducing cell cycle-associated genes in breast cancer cells. Mol Cancer Res 1:300–311

Howell A, Cuzick J, Baum M, Buzdar A, Dowsett M, Forbes JF, Hoctin-Boes G, Houghton J, Locker GY, Tobias JS (2005) Results of the ATAC (Arimidex, Tamoxifen, Alone or in Combination) trial after completion of 5 years' adjuvant treatment for breast cancer. Lancet 365:60–62

Hui R, Finney G, Carroll JS, Lee CSL, Musgrove EA, Sutherland RL (2002) Constitutive overexpression of cyclin D1 but not cyclin E confers acute resistance to antiestrogens in T-47D breast cancer cells. Cancer Res 62:6916–6923

Kilker RL, Hartl MW, Rutherford TM, Planas-Silva MD (2004) Cyclin D1 expression is dependent on estrogen receptor function in tamoxifen-resistant breast cancer cells. J Steroid Biochem Mol Biol 92:63–71

Hunter T, Pines J (1994) Cyclins and cancer. II: Cyclin D and CDK inhibitors come of age. Cell 79:573–582

Koff A, Giordano A, Desai D, Yamashita K, Harper JW, Elledge S, Nishimoto T, Morgan DO, Franza BR, Roberts JM (1992) Formation and activation of a cyclin E-cdk2 complex during the G1 phase of the human cell cycle. Science 257:1689–1694

Lai A, Sarcevic B, Prall OW, Sutherland RL (2001) Insulin/insulin-like growth factor-I and estrogen cooperate to stimulate cyclin E-Cdk2 activation and cell Cycle progression in MCF-7 breast cancer cells through differential regulation of cyclin E and p21$^{WAF1/Cip1}$. J Biol Chem 276:25823–25833

Lukas J, Bartkova J, Bartek J (1996) Convergence of mitogenic signalling cascades from diverse classes of receptors at the cyclin D-cyclin-dependent kinase-pRb-controlled G1 checkpoint. Mol Cell Biol 16:6917–6925

McDonnell DP, Norris JD (2002) Connections and regulation of the human estrogen receptor. Science 296:1642–1644

Mukherjee S, Conrad SE (2005) c-Myc suppresses p21WAF1/CIP1 expression during estrogen signaling and antiestrogen resistance in human breast cancer cells. J Biol Chem 280:17617–17625

Musgrove EA, Wakeling AE, Sutherland RL (1989) Points of action of estrogen antagonists and a calmodulin antagonist within the MCF-7 human breast cancer cell cycle. Cancer Res 49:2398–2404

Musgrove EA, Hamilton JA, Lee CS, Sweeney KJ, Watts CK, Sutherland RL (1993) Growth factor, steroid, and steroid antagonist regulation of cyclin gene expression associated with changes in T-47D human breast cancer cell cycle progression. Mol Cell Biol 13:3577–3587

Musgrove EA, Hunter LJ, Lee CS, Swarbrick A, Hui R, Sutherland RL (2001) Cyclin D1 overexpression induces progestin resistance in T-47D breast cancer cells despite p27(Kip1) association with cyclin E-Cdk2. J Biol Chem 276:47675–47683

Oster SK, Ho CS, Soucie EL, Penn LZ (2002) The myc oncogene: MarvelouslY Complex. Adv Cancer Res 84:81–154

Planas-Silva MD, Weinberg RA (1997) Estrogen-dependent cyclin E-cdk2 activation through p21 redistribution. Mol Cell Biol 17:4059–4069

Planas-Silva MD, Donaher JL, Weinberg RA (1999) Functional activity of ectopically expressed estrogen receptor is not sufficient for estrogen-mediated cyclin D1 expression. Cancer Res 59:4788–4792

Prall OW, Sarcevic B, Musgrove EA, Watts CK, Sutherland RL (1997) Estrogen-induced activation of Cdk4 and Cdk2 during G1-S phase progression is accompanied by increased cyclin D1 expression and decreased cyclin-dependent kinase inhibitor association with cyclin E-Cdk2. J Biol Chem 272:10882–10894

Prall OW, Rogan EM, Musgrove EA, Watts CK, Sutherland RL (1998) c-Myc or cyclin D1 mimics estrogen effects on cyclin E-Cdk2 activation and cell cycle reentry. Mol Cell Biol 18:4499–4508

Prall OW, Carroll JS, Sutherland RL (2001) A low abundance pool of nascent p21WAF1/Cip1 is targeted by estrogen to activate cyclin E*Cdk2. J Biol Chem 276:45433–45442

Roy PG, Thompson AM (2006) Cyclin D1 and breast cancer. Breast 15:718–727

Rudolph B, Saffrich R, Zwicker J, Henglein B, Muller R, Ansorge W, Eilers M (1996) Activation of cyclin-dependent kinases by Myc mediates induction of cyclin A, but not apoptosis. EMBO J 15:3065–3076

Sabbah M, Courilleau D, Mester J, Redeuilh G (1999) Estrogen induction of the cyclin D1 promoter: involvement of a cAMP response-like element. Proc Natl Acad Sci USA 96:11217–11222

Sherr CJ, Roberts JM (1999) CDK inhibitors: positive and negative regulators of G1-phase progression. Genes Dev 13:1501–1512

Shiyanov P, Bagchi S, Adami G, Kokontis J, Hay N, Arroyo M, Morozov A, Raychaudhuri P (1996) p21 Disrupts the interaction between cdk2 and the E2F-p130 complex. Mol Cell Biol 16:737–744

Smyth GK, Michaud J, Scott HS (2005) Use of within-array replicate spots for assessing differential expression in microarray experiments. Bioinformatics 21:2067–2075

Sutherland RL, Reddel RR, Green MD (1983a) Effects of oestrogens on cell proliferation and cell cycle kinetics. A hypothesis on the cell cycle effects of antioestrogens. Eur J Cancer Clin Oncol 19:307–318

Sutherland RL, Hall RE, Taylor IW (1983b) Cell proliferation kinetics of MCF-7 human mammary carcinoma cells in culture and effects of tamoxifen on exponentially growing and plateau-phase cells. Cancer Res 43:3998–4006

Sutherland RL, Prall OW, Watts CK, Musgrove EA (1998) Estrogen and progestin regulation of cell cycle progression. J Mammary Gland Biol Neoplasia 3:63–72

Tang S, Han H, Bajic VB (2004) ERGDB: Estrogen Responsive Genes Database. Nucleic Acids Res 32:D533–536

Taylor IW, Hodson PJ, Green MD, Sutherland RL (1983) Effects of tamoxifen on cell cycle progression of synchronous MCF-7 human mammary carcinoma cells. Cancer Res 43:4007–4010

Venditti M, Iwasiow B, Orr FW, Shiu RP (2002) C-myc gene expression alone is sufficient to confer resistance to antiestrogen in human breast cancer cells. Int J Cancer 99:35–42

Wang L, Shao ZM (2006) Cyclin E expression and prognosis in breast cancer patients: a meta-analysis of published studies. Cancer Invest 24:581–587

Wakeling AE, Bowler J (1988) Novel antioestrogens without partial agonist activity. J Steroid Biochem 31:645–653

Watson PH, Pon RT, Shiu RP (1991) Inhibition of c-myc expression by phosphorothioate antisense oligonucleotide identifies a critical role for c-myc in the growth of human breast cancer. Cancer Res 51:3996–4000

Watts CK, Brady A, Sarcevic B, de Fazio A, Musgrove EA, Sutherland RL (1995) Antiestrogen inhibition of cell cycle progression in breast cancer cells in associated with inhibition of cyclin-dependent kinase activity and decreased retinoblastoma protein phosphorylation. Mol Endocrinol 9:1804–1813

Watts CK, Prall OWJ, Carroll JS, Wilcken NRC, Rogan EM, Musgrove EA, Sutherland RL (2002) Antiestrogens and the Cell Cycle. In: Jordan VC, Furr BJ (eds) Antioestrogens and Antiandrogens. Humana Press, Totowa, New Jersey, pp 17–45

Weinberg RA (1995) The retinoblastoma protein and cell cycle control. Cell 81:323–330

Wong CW, McNally C, Nickbarg E, Komm BS, Cheskis BJ (2002) Estrogen receptor-interacting protein that modulates its nongenomic activity-crosstalk with Src/Erk phosphorylation cascade. Proc Natl Acad Sci USA 99:14783–14788

Wilcken NR, Prall OW, Musgrove EA, Sutherland RL (1997) Inducible overexpression of cyclin D1 in breast cancer cells reverses the growth-inhibitory effects of antiestrogens. Clin Cancer Res 3:849–854

Zeller KI, Jegga AG, Aronow BJ, O'Donnell KA, Dang CV (2003) An integrated database of genes responsive to the Myc oncogenic transcription factor: identification of direct genomic targets. Genome Biol 4:R69

Zhu L, Harlow E, Dynlacht BD (1995) p107 uses a p21CIP1-related domain to bind cyclin/cdk2 and regulate interactions with E2F. Genes Dev 9:1740–1752

Previous titles in the book series Research and Perspectives in Endocrine Interactions

C. Kordon, I. Robinson, J. Hanoune, R. Dantzer, Y. Christen (eds) (2003)
Brain Somatic Cross-Talk and the Central Control of Metabolism
ISBN 978-3-540-00090-7

P. Chanson, J. Epelbaum, S. Lamberts, Y. Christen (eds) (2004)
Endocrine Aspects of Successful Aging: Genes, Hormones and Lifestyles
ISBN 978-3-540-40573-3

C. Kordon, R.C. Gaillard, Y. Christen (eds) (2005)
Hormones and the Brain
ISBN 978-3-540-21355-0

J.-C. Carel, P.A. Kelly, Y. Christen (eds) (2005)
Deciphering Growth
ISBN 978-3-540-26192-6

M. Conn, C. Kordon, Y. Christen (eds) (2006)
Insights into Receptor Function and New Drug Development Targets
ISBN 978-3-540-34446-9